BEETLES
OF THE WORLD

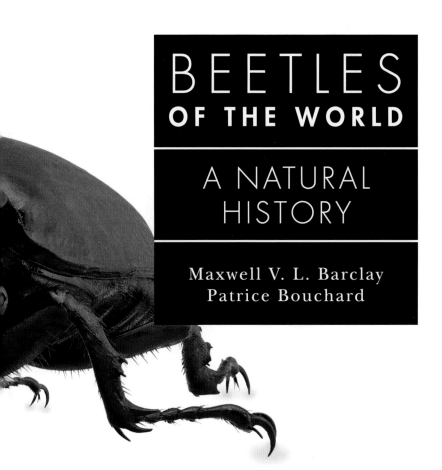

BEETLES
OF THE WORLD

A NATURAL HISTORY

Maxwell V. L. Barclay
Patrice Bouchard

PRINCETON UNIVERSITY PRESS
PRINCETON AND OXFORD

Published in 2023 by Princeton University Press
41 William Street, Princeton, New Jersey 08540
99 Banbury Road, Oxford OX2 6JX
press.princeton.edu

Conceived, designed, and produced by
The Bright Press
an imprint of The Quarto Group
1 Triptych Place, London, SE1 9SH, United Kingdom
T (0) 20 7700 6700
www.quarto.com

Library of Congress Control Number: 2022951063
ISBN: 978-0-69124-073-2
Ebook ISBN: 978-0-69124-074-9
British Library Cataloging-in-Publication Data is available

Publisher James Evans
Editorial Director Isheeta Mustafi
Art Director James Lawrence
Managing Editor Jacqui Sayers
Senior Editors Caroline Elliker, Sara Harper, Izzie Hewitt
Project Editor Katie Crous
Design Kevin Knight
Picture Research Sharon Dortenzio
Illustrations John Woodcock

Cover and prelim photos: Front cover (clockwise from top left): *Leptinotarsa decemlineata* (Chrysomelidae), *Cryptocephalus sericeus* (Chrysomelidae), *Mormolyce phyllodes* (Carabidae), *Anthaxia candens* (Buprestidae), *Meloe proscarabaeus* (Meloidae), *Diaperis boleti* (Tenebrionidae) (top), *Dytiscus marginalis* (Dytiscidae) (bottom), *Malachius bipustulatus* (Melyridae), *Sagra buqueti* (Chrysomelidae). A photo of each of these species in their natural habitat is included in this book.
Spine: *Harmonia axyridis* (Coccinellidae)
Back cover: *Lamprima aurata* (Lucanidae)
Page 2–3: *Eupatorus birmanicus* (Scarabaeidae)
Page 5: *Coccinella septempunctata* (Coccinellidae)

Printed in Malaysia

10 9 8 7 6 5 4 3 2 1

6 Introduction

CONTENTS

82 Taxonomy

82 Archostemata

86 Myxophaga

90 Adephaga

108 Polyphaga

228 Glossary
230 Resources
231 About the Authors
232 Index
239 Picture Credits
240 Acknowledgments

INTRODUCTION

Beetles (order: Coleoptera) are arguably the most diverse and species-rich group of organisms on the planet. Almost 400,000 species have been formally described so far, more than double any equivalent group. Thousands more are discovered each year, and beetles comprise around 25 percent of known animal life on earth. Estimates suggest that even these incredible numbers just scratch the surface of what is actually out there in the tropical forests of the world, and there is a great urgency to document, understand, and archive this extraordinary diversity while the habitats that support it still exist and remain accessible for study.

There is a quotation that coleopterists like, from the evolutionary biologist J.B.S. Haldane (1892–1964), who was apparently asked by a group of clerics if his studies of the natural world had taught him anything about the mind of God. He is said to have replied that if there is a Creator, He must have had "an inordinate fondness for beetles." The use of the word "inordinate" here is interesting, as it is not entirely positive; it suggests that this supposed divine fondness, or at least the number of beetles in the world, is disproportionate or even inappropriate—but Haldane was not an entomologist and did not fully understand. In fact, the extraordinary diversity of beetles has produced an unintended consequence, an iterative and truly enormous benefit not just for Haldane and his generation of geneticists, but also for the entire human race. Beetle diversity helped humanity to solve the biggest question of all, the question of our place in nature, why we are here. This may sound like a wild claim, but both

LEFT | *Chrysochroa fulgidissima* (Buprestidae)
A Metallic Wood Boring Beetle from the forests
of eastern Asia. Adults fly in sunshine, and the
stunning colors are caused by refraction of light.

Charles Darwin (1809–82) and Alfred Russel Wallace (1823–1913), the two Victorian naturalists who independently disentangled and discovered the mechanism of natural selection, the driving force of evolution, were by hobby and inclination beetle collectors. People who immerse themselves in the meticulous study of such hyperdiverse organisms may be seeing a picture of the world with a lot more "pixels," a lot more detail. During detailed study and identification of beetles, huge similarities and minute differences between species, which inhabit different areas or even have completely different behaviors, make it difficult to escape the thought that they may in fact be related, and have a common origin. The observer may start to ponder the reasons for the changes. This provided the spark, the eureka moment for both Darwin and Wallace, and the rest fell into place. It is a humbling thought that human understanding of our place in the universe is partly because two young people, 150 years ago, liked collecting beetles.

Beetles dominate and maintain terrestrial ecosystems worldwide, being absent only from the oceans and Antarctica. Almost anywhere on the surface of the world, one is never more than a few feet from a beetle. It is hoped that this book will give some overview of their vast range of form, function, and behavior, and will help to stimulate a curiosity and fascination with this essential, hyperdiverse, and truly inspiring group of animals.

WHAT ARE BEETLES?

Alfred Russel Wallace, the famous evolutionary biologist, biogeographer, and coleopterist, reflected on the question, "What are beetles?"

"It is a melancholy fact that many of our fellow creatures do not know what is a beetle! They think cockroaches are beetles! Tell them that beetles are more numerous, more varied, and even more beautiful than the birds or beasts or fishes that inhabit the earth and they will hardly believe you; tell them that he who does not know something about beetles misses a never failing source of pleasure and occupation and is ignorant of one of the most important groups of animals inhabiting the earth, and they will think you are joking; tell them further that he who has never observed and studied beetles passes over more wonders on every field and every copse than the ordinary traveler sees who goes round the world and they will perhaps consider you crazy, yet you will have told them only the truth."

One hopes that awareness of beetles and their vast diversity, myriad forms and functions, and ecological importance is greater today than 150 years ago, when Wallace wrote these lines—although confusion with cockroaches continues!

So what is a beetle? It is an insect of the order Coleoptera. As an insect, the adult body is divided into three parts: head, thorax, and abdomen, and it has six legs. Coleoptera are defined by the front pair of wings modified into elytra, which protect the abdomen and the folded flight wings.

LEFT | *Golofa porteri*
(Scarabaeidae: Dynastinae)
The male of this South American rhinoceros beetle has some of the most extreme horns of any insect, but despite this is able to fly.

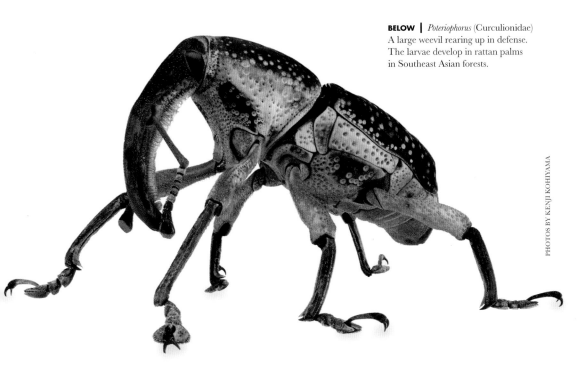

BELOW | *Poteriophorus* (Curculionidae) A large weevil rearing up in defense. The larvae develop in rattan palms in Southeast Asian forests.

PHOTOS BY KENJI KOHIYAMA

Beetles have biting mouthparts, distinguishing them from true bugs (Heteroptera), which have a sucking tube. The key difference between beetles and superficially similar insects such as bugs and cockroaches is the fact that Coleoptera have complete metamorphosis, that is they belong to the division of insects called Holometabola, which means that they start as an egg that develops into a larva, a feeding stage, which then becomes a pupa, out of which emerges the adult, with its wings and reproductive organs. Cockroaches and true bugs, both belonging to Hemimetabola, have immature stages that resemble smaller wingless versions of the adult, and they never have a pupal stage.

From a taxonomic point of view, beetles are well defined, and apart from some Carboniferous fossils it is usually clear what is and what is not a beetle. The only exception is the small order Strepsiptera, called twisted-winged parasites. Males of these small insects have only the hindwings developed for flight. The rod-like forewings seem to serve a balancing function. The female is parasitic, eyeless, and limbless, and lives between the abdominal segments of another insect, usually a bee or wasp. Only the abdomen is visible, so that the males can mate. Strepsiptera are often considered the closest living relatives (Sister Group) of beetles, but some scientists have good reason to consider that they belong well inside the beetles, in Tenebrionoidea.

BEETLE ANATOMY

As in all insects, the body of an adult beetle is divided into three distinct sections: the head, thorax, and abdomen. The head is where the mouthparts and the main sense organs (the compound eyes and antennae) are based. The thorax is the second division of the insect body, between the head and the abdomen. In beetles, the part of the thorax visible from above is the pronotum, and this is often informally called the thorax, but the whole thorax includes two more parts, the mesonotum and metanotum, which are usually covered by the elytra. This means they are not visible from above, but can be seen on the underside of the beetle or when the wings and elytra are open for flight. The abdomen is the third and final of the three sections of the insect body, and attaches to the metathorax just after the attachment point of the hind pair of legs.

BODY AND STRUCTURE

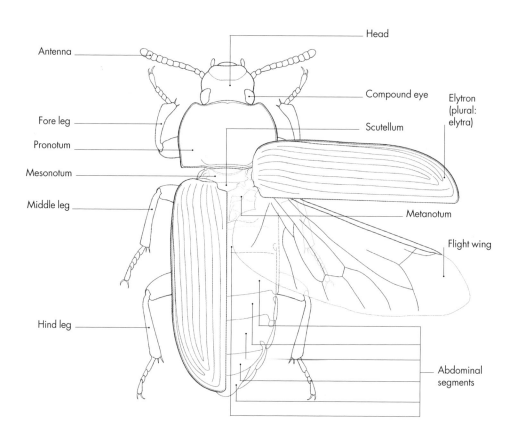

Antenna

Fore leg

Pronotum

Mesonotum

Middle leg

Hind leg

Head

Compound eye

Scutellum

Elytron (plural: elytra)

Metanotum

Flight wing

Abdominal segments

INTERNAL ANATOMY

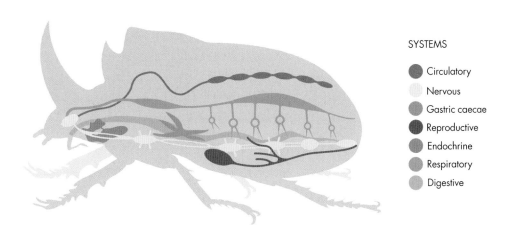

SYSTEMS

- ● Circulatory
- ○ Nervous
- ● Gastric caecae
- ● Reproductive
- ● Endochrine
- ● Respiratory
- ● Digestive

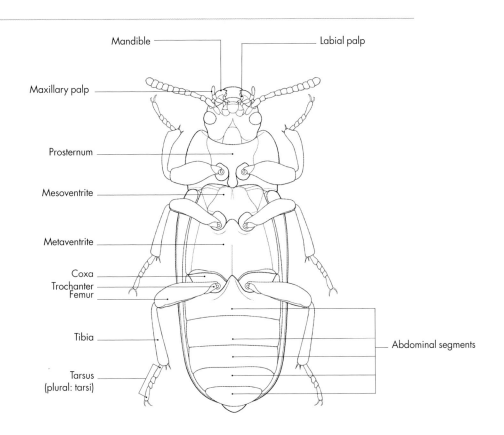

Mandible

Labial palp

Maxillary palp

Prosternum

Mesoventrite

Metaventrite

Coxa
Trochanter
Femur

Tibia

Abdominal segments

Tarsus
(plural: tarsi)

HEAD

Adult beetles have two large compound eyes, except where they have been lost, for example in numerous myrmecophilous (ant-associated), subterranean, or troglobitic (cave-dwelling) species of Carabidae, Dytiscidae, Staphylinidae, Leiodidae, and so on, which inhabit near-permanent darkness, and have no need for vision. Many other beetles, especially those that are active during the day, have excellent eyesight. Many flower-feeding beetles, for example, can home in on blossom of a particular kind and color from high in the air, while jewel beetles (Buprestidae) are extremely visually alert to the approach of predators and are almost impossible to catch, even with a net. Tiger beetles (Carabidae: Cicindelinae) are visual predators and some are quick enough to grab flies from the air. Diurnal dung-inhabiting Staphylinidae of the genus *Ontholestes* are also notable for their ability to grab adult flies, using a combination of excellent vision and speed. Many aquatic beetles use their compound eyes in water, and the eyes of Gyrinidae (whirligig beetles) are divided so that they can use the upper half to scan the air above, while the lower half observes the water below.

Unlike many insect orders, very few beetles have secondary eyes, or ocelli, as adults. Most Dermestidae (hide beetles) have a single ocellus in the center of the head, which is a useful identification character, and some Staphylinidae have ocelli.

The mandibles, or jaws, are also attached to the head, surrounding the mouth, with two sets

BELOW | *Manticora scabra* (Carabidae) Named after a mythological monster, this male African tiger beetle uses asymmetrical mandibles for crushing large prey and grasping a mate.

of limb-like sensory organs called maxillary and labial palps, which taste and process food. Mandibles may be only for eating, or they may be developed for hunting and defense, as in many Carabidae, or for display, as in male stag beetles (Lucanidae). In most weevils (Curculionoidea) and a few other families (some Lycidae, Salpingidae) the mandibles are placed on a beaklike extension of the head called a rostrum.

Probably the most important sensory structures of the head in beetles are the antennae. We do not have a directly equivalent sense, and we often refer to them as "feelers," but the sense that they provide is much closer to smell than to touch. Many beetle antennae, especially in males, are lamellate or pectinate to maximize surface area so that they can detect small quantities of chemicals in the air, for example pheromones produced by the female. The importance of antennae is shown by the fact that many adult beetles have lost their eyes, but none have lost their antennae. The original beetle ground plan includes 11 segmented

antennae, but in a few families and genera some segments have been lost, or in a few cases, some more gained. The shape and structure are important taxonomically.

Finally, in some beetles, for example in Scarabaeidae, the head has horns, which may connect with horns on the thorax.

THORAX

All six legs, as well as the elytra and flight wings, are attached to the thorax (see diagram on page 10), and this holds the muscles to operate these limbs. The intestine passes through the center, but most of the thorax consists of muscle.

The exoskeleton of the thorax, to which these powerful muscles are attached from the inside, needs to be strong, and the upper surface of the pronotum is usually the hardest part of the beetle. It may have horns, for example in the Hercules Beetle *Dynastes hercules* (Scarabaeidae: Dynastinae) the forward-pointing thoracic horn is much longer than the head (or cephalic) horn, and protects the head from above. A few weevils in the subfamily Baridinae have forward-pointing horns, or tusks, on the underside of the thorax, which the males use in fighting for females; each male places its horns into special sockets in the thorax of the other male, and then they have a trial of strength where they try to push one another off a twig.

Many other beetles that are not horned, such as some Tenebrionidae (*Cossyphus rugosulus*, illustrated opposite below) and many species of Lampyridae, Cantharidae, Chrysomelidae, and Bostrichidae, have part of the thorax shielding the head from above, so that the "front" of the beetle is the front margin of the thorax, with the head below and behind. Some have pale spots in the thorax corresponding to where the eyes are, so that the beetle can still see changes in light.

The extended thorax protects the head, eyes, and antennae, but it may also have other uses. For example, in male fireflies (Lampyridae), which fly in search of the light produced by a female, the projection of the thorax shields the eyes from above, like the blinkers worn by a horse. This prevents the beetle from being distracted by the stars, which might otherwise appear brighter than the glowing female they are looking for.

LEFT | *Cyclommatus asahinai* (Lucanidae) This Oriental Stag Beetle is taking flight, showing the rear parts of the thorax that are usually covered by the elytra.

OPPOSITE ABOVE | *Theodosia rodorigezi* (Scarabaeidae) A male of this Philippine flower chafer shows cephalic and thoracic horns.

OPPOSITE BELOW | *Cossyphus rugosulus* (Tenebrionidae) This Mediterranean darkling beetle has the thorax covering the head; pale spots let some light through to the eyes.

Many families, such as Dermestidae, Histeridae, and Byrrhidae, have grooves in the thorax into which the antennae or limbs are withdrawn for protection when not in use. Many weevils, especially the tribe Cryptorrhynchini, have a groove in the center of the underside of the thorax to accept the rostrum. The thorax is often decorated; for example, in the common Seven Spot Ladybird *Coccinella septempunctata* (Coccinellidae), the thorax has white "eye spots" that are much larger than the beetle's actual eyes. In click beetles of the Pyrophorini (Elateridae), the bright bioluminescent "headlamps," probably also false eyes, are on the upper surface of the pronotum.

ABDOMEN

The abdomen (see diagram on page 11) consists of segments called tergites, five to nine of which are generally visible (others being withdrawn into the apex, and able to extend during mating or egg laying). Each tergite has a pair of spiracles, external openings of the respiratory system that allow air into the trachea from which oxygen is diffused into the tissues.

In a typical beetle at rest, the abdomen is covered by the folded flight wings and the closed elytra, though the apical few segments, called a pygidium, may extend beyond the elytra. In some families, especially the rove beetles (Staphylinidae), but also some Nitidulidae, Cantharidae, and other families, the elytra are short, leaving several abdominal segments exposed. This makes them more flexible and able to move through narrow tunnels or dense substrate.

At the apex, that is the tip, of the abdomen are the anus and the genitalia. The flexibility of the apical segments of the abdomen is important in mating in many beetles, since many parts of the exoskeleton are rigid and not freely movable. The male genitalia of many families of beetles are also important for taxonomy, to distinguish between similar species. When the external morphology is uniform, the structure of the genitalia can still be different, often as a deliberate evolutionary strategy to prevent male beetles from accidentally mating with females of similar but distinct species that occur in the same environment. Mating will not proceed if the genitalia are incompatible, because mating with an incorrect species may result in no offspring or offspring that are infertile. This is known as the "lock and key" hypothesis. Scientists often routinely dissect the genitalia when preparing specimens of beetles of difficult groups.

Some species are called physogastric (swollen bellied), as the abdomen in the female swells with eggs, making flight impossible. This is particularly common in some leaf beetles (Chrysomelidae), such as the common European Green Dock Beetle *Gastrophysa viridula*.

Beyond defecation and reproduction, the abdomen of beetles serves a variety of other functions. It is where the bioluminescence is produced in Lampyridae, for example, in a light

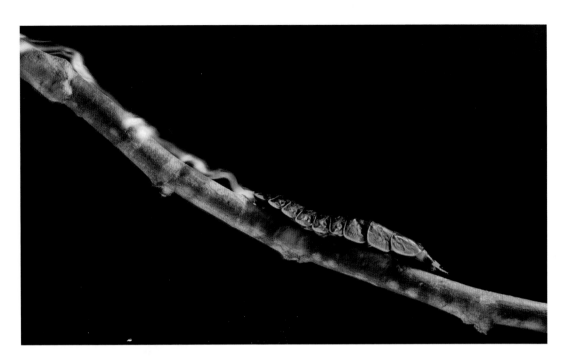

organ on the underside of several apical
abdominal segments. It is also used to produce
and distribute defensive chemicals. Staphylinidae
have a stink organ at the apex of the abdomen,
and larger species such as the Devil's Coach Horse
Ocypus olens raise the abdomen into a threat
position over their head, making them look
scorpion-like, while at the same time producing
a defensive smell. Some Carabidae,
particularly the bombardier beetles of
the genus *Brachinus*, have taken chemical
defense to an extreme, and have an
explosion chamber at the apex of the
abdomen, where, when alarmed, they
mix glandular secretions that allow
them to shoot a scalding hot cocktail
of chemicals at an attacker.

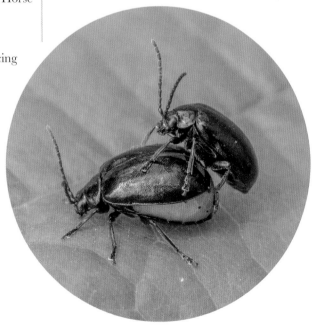

EXAGGERATED STRUCTURES

The Asian Atlas Beetle *Chalcosoma* (Scarabaeidae: Dynastinae) is a rhinoceros beetle where the male has three long, curved horns, one on the head and two on the thorax, and is a glossy metallic color. Males in the breeding season are conspicuous, flying around streetlights in tropical towns in Malaysia or Indonesia, and it is surprising that they can fly at all with their extreme ornamentation. If two males meet, they immediately compete, sizing off against each other, or grappling and pushing each other with their moveable cephalic horns until one is defeated. Although it impressed Darwin, and would impress most observers, the purpose of this extravagant display is to impress the female *Chalcosoma*, which is a dull brownish color and lacks any horns, is rarely seen, and does not display, but instead maintains a low profile among tangled vegetation. The females safeguard the future of the species, as the majority of males will never reproduce and, for all their extravagance, are the expendable sex. The extreme decoration of the males is also maintained and passed to the next generation by the females, since females preferentially reproduce

BELOW | *Phanaeus demon* (Scarabaeidae) A suitably demonic-looking male of the Central American rainbow dung beetle showcases many exaggerated structures.

BELOW | *Sagra buqueti* (Chrysomelidae) The Frog-Legged Beetle from tropical Asia is, paradoxically, unable to hop with its huge back legs.

OPPOSITE | *Cyclommatus eximius* (Lucanidae) The male of this New Guinean stag beetle has huge mandibles. The female has no such ornamentation.

with those males with the greatest ornamentation, ensuring that their own male offspring receive the genes for the maximum showy horns and bright colors, and that the pageant continues.

The situation in *Chalcosoma* is a classic example of sexual selection, where a characteristic with no obvious value for survival—in fact with negative survival cost, as it is expensive to produce in terms of resources, and makes the bearer less agile, more conspicuous, and at greater risk of predation—is maintained and enhanced because its possessors are, seemingly arbitrarily, more appropriate mates. Sexual selection is at odds with regular natural selection, in this case the former driving the extreme adaptations of the male; the latter favoring the camouflaged, secretive, unadorned female as a better survival strategy. Parallel situations exist with the tails of peacocks and

some pheasants, or the antlers of deer, or, some might add, certain expensive sports cars.

The adoption of extreme structures via sexual selection appears throughout the Coleoptera but is most obvious in the Scarabaeoidea, and it is not necessary to go to tropical Asia to observe this. Stag beetles (Lucanidae) in gardens and yards in Europe and North America demonstrate the same effect. Not all extreme modifications result from sexual selection. The huge, fanlike antennae of some Elateroidea serve a practical function, to sift the air for minute quantities of target chemicals. The huge hind legs of some Chrysomelidae, subfamilies Bruchinae and Sagrinae (see *Sagra buqueti*, left), may serve an antipredation function, which is still not fully understood.

BEETLE LIFE CYCLE

Beetles are holometabolous insects, which means that they have complete metamorphosis, an extraordinarily complex but successful adaptation that has probably only evolved once in the history of life. Other such insects include the other three hyperdiverse insect orders, Diptera (true flies), Hymenoptera (bees, ants, and wasps) and Lepidoptera (butterflies and moths). These four orders combined total over 820,000 named species, so together they account for considerably more than half of all known animal life, suggesting that complete metamorphosis is correlated with astonishing diversity.

So, what is complete metamorphosis? Most people are familiar with the life cycle of a butterfly, and the basic life cycle of a beetle or any other holometabolous insect is essentially the same. The female lays fertilized eggs, which hatch into larvae (also called, depending on the group, grubs, maggots, or caterpillars). These are a feeding stage and have limited sense organs and limited mobility. Some are blind, others legless, and beetle larvae consume a wide range of substances, but many live surrounded by food and do nothing except eat and grow. As they grow they molt, casting off their old cuticle and inflating the new, soft one beneath with air to give them space to grow until the next molt, a process called "ecdysis." After a number of molts, the fully grown larva searches for a place to pupate. The pupa is a quiescent stage, hardly able to move, so unable to escape from predators, parasites, dehydration, or flooding, making pupation a vulnerable period in the insect's development. Many beetles pupate deep inside substrate, making a protected pupal cell in the soil or in dead wood. Inside the pupa, the internal organs

BELOW LEFT | *Hypera rumicis* (Curculionidae) The latticed pupal cases of this externally feeding weevil in England, UK, give some protection from predators and parasitoids.

BELOW RIGHT | *Omaspides* (Chrysomelidae: Cassidinae) A tortoise beetle from Ecuador shows parental care of a brood of more than 30 larvae. Larvae retain their excrement, as a physical and chemical defense.

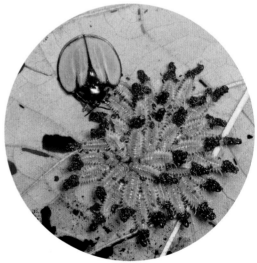

FROM EGG TO BEETLE

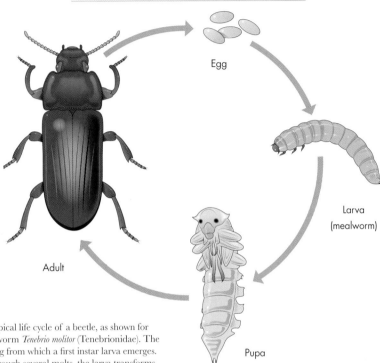

Egg

Larva
(mealworm)

Pupa

Adult

ABOVE | The typical life cycle of a beetle, as shown for the Yellow Mealworm *Tenebrio molitor* (Tenebrionidae). The female lays an egg from which a first instar larva emerges. After growing through several molts, the larva transforms into a pupa. The adult male and female beetles emerge from the relatively short pupal stage and start the cycle again.

of the larvae liquify and reform into a completely different looking insect, the adult. When the adult is fully formed, it splits open the pupal shell and emerges, at which point its exoskeleton is still soft and whitish colored, so it is unable to fly, a condition referred to as "teneral." Over a number of hours or days, the cuticle and exoskeleton harden, allowing the insect to begin its adult life.

This life cycle— egg, larva, pupa, adult— contrasts with that of the older and simpler insect orders, called hemimetabolous insects. These include the Orthoptera (grasshoppers and crickets), Blattodea (cockroaches and termites), Hemiptera (true bugs), and numerous other

groups, which have a life cycle where the egg hatches into a nymph (this looks like a smaller, wingless version of the adult). The nymph then grows by molting until it eventually becomes an adult, with no pupal stage. One of the advantages of the holometabolous system is that the larva is totally different from the adult, so it often does not live in the same habitat, does not eat the same food, and is not subject to the same dangers. This reduces competition between generations of the same species, and uses the resources of the environment more efficiently, and it may also be one of the factors that has led to the success of the "big four" insect orders.

BEETLE FOSSIL HISTORY

The oldest undisputed beetle fossils date from the early Permian, about 290 million years ago. They belong to the extinct taxon Protocoleoptera, usually treated as an extinct suborder that is sister group (closest relative) to all remaining Coleoptera. Most fossils are between 5 and 20 mm; some are well-preserved impressions of complete insects, while most are represented only by elytra. They show that Protocoleoptera are beetles, although they have unusual features. The female has an ovipositor (egg-laying tube) which is unpaired and apically pointed, and nothing similar exists in any modern beetles. The elytra have a conspicuous network of veins,

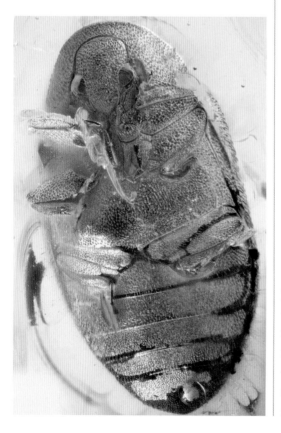

are pointed apically, and extend beyond the tip of the abdomen. The extinct Permian families of Protocoleoptera include Tshecardocoleidae, Oboricoleidae, and Moravocoleidae, believed to have been wood feeders as larvae, and considered to be similar to the extant Archostemata. Even older fossils, such as *Adiphlebia lacoana* from the Carboniferous period of North America, have been interpreted by some authors as beetles, but this is not widely accepted.

As they have robust exoskeletons, adult beetles have a better chance of preservation in the fossil record than many insects. All four modern suborders—Archostemata, Myxophaga, Adephaga, and Polyphaga—are well known in the fossil record, as are more than 60 percent of modern families. Those that are missing are often small families, or associated with habitats where preservation is unlikely.

The beetle fossil record was given an enormous boost by the proliferation in the Cretaceous of amber-producing trees, which produce a sticky sap in which insects and other animals and plant fragments become trapped, and which then hardens into resin and ultimately fossilizes into amber. Insect fossils in amber, called "inclusions," are often perfectly preserved in three dimensions, and skillful polishing can reveal the insects from the desired angles. The superior preservation makes the study of amber fossils easier and less subjective than the study of stone fossils. While the most familiar ambers are the northern European Baltic amber (35 to 48 million years old) and Caribbean Dominican amber (25 million years old), these are both Cenozoic and include a fauna not very different at the family, and sometimes the genus level, from that found today. Mesozoic amber opens a door to an earlier and more different fauna. Cretaceous Burmese amber from Myanmar is about 99 million years old, while Cretaceous Lebanese amber is older, at 125 to 135 million years. Some Jurassic Lebanese amber with beetle inclusions is also known.

Recent breakthroughs in synchrotron technology have allowed scientists to digitally reconstruct insects preserved in nontransparent amber. The importance of this opaque amber was never before realized, because the inclusions were not visible.

The chances of any individual becoming a fossil are vanishingly small, and what we know about the fauna of the past relies on chance preservation in the fossil record. It is an important role of modern entomologists to build insect collections, as a new fossil record, so future generations will have a more complete window into the insect fauna of the world today.

RIGHT | *Pelretes vivificus* (Kateretidae) A pollinating beetle from mid-Cretaceous Burmese amber, provisionally assigned to the extant family Kateretidae.

OPPOSITE | *Chelonarium andabata* (Chelonariidae) One of two fossil turtle beetles from Eocene Baltic amber, assigned to a modern, mainly tropical, genus.

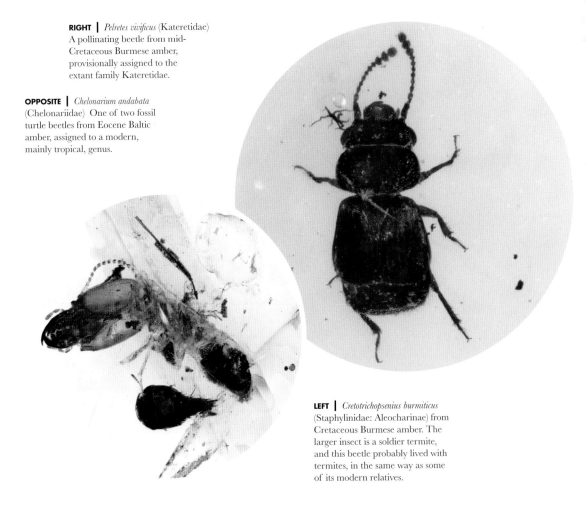

LEFT | *Cretotrichopsenius burmiticus* (Staphylinidae: Aleocharinae) from Cretaceous Burmese amber. The larger insect is a soldier termite, and this beetle probably lived with termites, in the same way as some of its modern relatives.

BEETLE FEEDING HABITS

PREDATORS

Predation, feeding on other animals, is a common feeding strategy for beetles. Almost all the suborder Adephaga, which includes the ground beetles (Carabidae) and diving beetles (Dytiscidae), are predatory, as are a large part of the suborder Polyphaga, for example most of the Coccinellidae (ladybugs), Staphylinidae (rove beetles), and Cleridae (checkered beetles). The large insect order most closely related to Coleoptera, the Neuroptera, which includes lacewings and antlions, are also almost all predators as adults and larvae. It is tempting to think it might be the ancestral state in beetles and their relatives, but the Archostemata, which are the best represented group in the ancient fossil record, and are thought to most resemble ancestral beetles, are wood and detritus feeders.

Here, predation is defined as the killing and consuming of other animals. Those beetles that eat meat which is already dead—for example, burying beetles (Silphidae) and many scarab beetles (Scarabaeoidea)—are classified as scavengers or necrophages.

Meat is a very high-energy food source, which is usually easy to metabolize, but first it needs to be caught. The adaptations of predators are therefore geared toward speed, strength, and stealth, instead of toward detoxifying and breaking down complex chemicals, as required by herbivores. While hunting for prey, predators are exposed to risks and enemies of their own, so efficiency is in their best interest, and targeting a few larger prey items is often more efficient than hunting down large numbers of small insects. The ground beetles (Carabidae) are a good example of active predators, with long legs adapted for running, good eyesight, and powerful forward-facing mandibles. Most of the temperate species hunt at night on the forest floor, attacking worms, slugs, and snails. Some carabids hunt faster-moving prey, and this is taken to an extreme

LEFT | *Carabus intricatus* (Carabidae) The Blue Ground Beetle, a rare species in Europe, is a nocturnal predator in woodlands, here feeding on an earthworm.

in the diurnal tiger beetles (Cicindelinae), which catch ants and flies; they include the fastest of all beetles and have exceptional vision as well as very strong, sharp mandibles. In the water, the diving beetles (Dytiscidae) fill the role of ground beetles as active hunters, feeding on leeches, water snails, and sometimes amphibians and fish.

In the suborder Polyphaga, which includes the majority of all beetles, many feeding strategies are used, and some of the most diverse groups have made the switch to living plants, but some families remain primarily predatory. Staphylinidae, as adults and larvae, are either carabid-like active hunters of invertebrates (such as the Devil's Coach Horse, *Ocypus olens*) or are predators in specific environments, for example hunting for fly larvae in dung or carrion, or ant larvae, while living as myrmecophiles in ant nests. Most Coccinellidae are predators of soft-bodied aphids or scale insects, which occur in such large numbers that adult or larval ladybugs hardly have to hunt at all, but simply graze on clusters of barely mobile prey. This apparently easy food supply comes at a cost, because the aphids and scale insects are exposed on the plants, making the ladybugs, in turn, exposed to other predators; to protect themselves the ladybugs secrete unpleasant chemicals, combined with bright warning coloration to deter their enemies.

LIVE PLANT TISSUE

Living plants, in forests, vegetated swamps, and grasslands, cover much of the world's land surface and shape terrestrial habitats; most of the nutrients in land ecosystems are locked up in living plants. Any animals that can adapt to feed on this vast resource can reap great ecological benefits. Unlike animals, plants cannot hide or run away; nor can they defend themselves with horns or jaws—but they are, nevertheless, difficult to eat for most animals. For a start, the cellulose of which they are built requires a complex intestine, usually including symbiotic bacteria, to extract any nutrition from it. Additionally, many plants produce toxic defensive compounds specifically to prevent insects and other herbivores from eating

them—in fact, humans extract some of these chemicals to make insecticides such as pyrethroids.

The relatively few groups of insects that have successfully circumvented plant defenses, either by metabolizing them or by storing them to use for their own protection, have become extremely species-rich and abundant as a result. These include butterflies and moths (Lepidoptera) and several superfamilies of beetles, particularly the families Chrysomelidae (Chrysomeloidea), the leaf beetles, and Curculionidae (Curculionoidea), the weevils, which are two of the largest groups in the whole animal kingdom, both exceeding 30,000 known species.

Plant-feeding, or phytophagous, insects are usually specialized to feed on a single genus or species of plant, which allows them to adapt to overcome the specific chemical protection of their target host plant. Over long periods of time, some plants evolve increasingly powerful toxins, and herbivores adapt by developing ever more effective means of detoxifying the compounds. This is called an "evolutionary arms race," and it increases the association between the insect and the host plant, since the specialized insects become the only herbivores able to feed on the toxic plant, and may even be attracted by the smell of the very chemicals evolved to combat them. The closeness of the association between phytophagous beetles and specific plants may have driven the diversification of beetles, since their ancestors were feeding on the ancestors of flowering plants, which diversified rapidly.

Phytophagous insects are also usually adapted to feed only on a particular structure of the host plant. So, for example, an oak tree may have

LEFT | *Scolytus multistriatus* (Curculionidae) Adults and larvae of Elm Bark Beetle feed on living trees, and can transmit the tree-killing Dutch elm disease fungus.

LEFT | *Eupholus geoffroyi* (Curculionidae)
Many exposed leaf feeders, such as this
weevil from New Guinea, use bright
warning colors that discourage predators.

BELOW | *Trachys minutus* (Buprestidae) This tiny European
jewel beetle mines between the laminae of a hazel leaf,
protected from predators and desiccation.

different beetles feeding on and developing in
its roots, stem, bark, twigs, leaves, buds, flowers,
and acorns.

Apart from the chemical defenses of the plants
themselves, plant feeders face other dangers.
Some plants have physical barriers such as sticky
resinous sap, which can gum up the mouthparts
of insects, or in some trees works like flypaper,
entrapping the whole insect (which can result in
specimens being preserved in fossilized resins,
such as amber, as discussed on page 22). Also,
phytophagous insects often need to feed while
openly exposed on the plant, making them
vulnerable to dehydration and predators. To avoid
this, some insects, at least as larvae, feed inside
the plant tissue, and those that feed openly have
evolved strategies, including complex crypsis and
camouflage, rapid jumping or flight reactions, or
chemical defenses of their own, to protect them.

DEAD AND DECAYING PLANT TISSUE

Plants make up most of the living biomass of the world's land ecosystems, and when they die they provide a vast food supply for other creatures, which break them down into nutrients used by other plants, so completing biological nutrient cycles. Compared to live vegetation, dead plants do not defend themselves with chemicals and are much more easily digested, so the range of beetles able to feed on them is much greater than when they were alive. Also, as soon as plants begin to decay they are invaded by huge populations of fungi, bacteria, and nematode worms, and many beetles that feed on dead plant matter are actually obtaining most of their nutrition from digesting other detritivores, rather than directly from the plant itself.

Judging from their closest living relatives, the larvae of the most ancient known beetles, from the suborder Archostemata, probably fed on dead plant matter, so dead plant tissue was the likely ancestral food of Coleoptera.

Dead plants should not be regarded as a single uniform resource; they consist of a whole progression from fermenting fruit to hard, dry timber. Vegetation decays in stages, from freshly dead and still green to humus and litter that is almost indistinguishable from soil, and all of these stages are utilized by different groups of beetles.

The hardest and most indigestible dead plant matter is the wood from the stems and branches of large trees, but beetle larvae of several families, notably Cerambycidae, Buprestidae, and Ptinidae, specialize in this and play an important role in breaking it down. If the wood becomes very dry, for example in a standing dead tree, or wood that has been made into furniture or structural timber, fungal and microbial activity almost shuts down, and very little nutrition remains for any beetle

RIGHT | *Lethrus apterus* (Geotrupidae) A flightless relative of dung beetles from Europe that collects dead leaves to feed to its larvae.

ABOVE | *Priacma serrata* (Cupedidae)
From the USA, this beetle develops
in dead wood. It is among the few
living representatives of the ancient
Archostemata.

BELOW | *Megasoma actaeon* (Scarabaeidae)
The heaviest known beetle, weighing
in at over 7 ounces (200 grams), is the
compost-feeding larva of a Central
American rhinoceros beetle.

larvae inside. Some larvae can survive even in these circumstances, but their development is greatly slowed down by the lack of nutrient content. In an extreme example, larvae of the North American jewel beetle *Buprestis aurulenta*, which typically last two to three years, emerged as adults from structural timber 25 years after it was harvested, dried, and processed!

On the opposite end of the decay spectrum, compost also consists mainly of rotting plant matter, but is very damp and filled with diverse communities of fungi, microbes, and invertebrates. This rich resource supports the largest known beetle larvae, the giant white grubs of some rhinoceros beetles, such as the South and Central American genera *Dynastes* and *Megasoma* (Scarabaeidae: Dynastinae). Many other scarabaeoid larvae develop in very moist and nutrient-rich decaying vegetation, from stag beetles in wet decayed subterranean wood to dung beetles—since dung is just dead vegetation that has passed through the gut of a vertebrate.

Based on comparison of related species, it is thought that the still-undiscovered larva of the biggest beetle of all, the South American Giant Longhorn *Titanus giganteus* (Cerambycidae), feeds on the decaying dead roots of giant trees under the soil, deep in the Amazonian jungles. It has still never been seen by scientists.

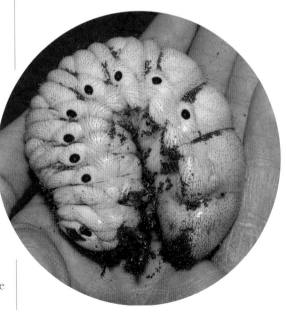

FUNGI

Fungi are incredibly diverse in morphology and ecology, ranging from microscopic single-celled yeasts to gigantic mushrooms; bread-molds to huge bracket fungi that grow for months or even years. Fungi show a very wide range of ecological relationships that involve Coleoptera, from beetle food to beetle parasites.

Many beetle families have evolved to exploit fungal fruiting bodies as a food resource, particularly abundant in wet tropical forests where fungi and beetles are among the most important

decomposers of wood and other vegetation. Specialized fungivores include Endomychidae (handsome fungus beetles), Erotylidae (pleasing fungus beetles), Ciidae (minute fungus beetles), Mycetophagidae (hairy fungus beetles), and many Tenebrionidae, particularly in the tribe Bolitophagini (Tenebrioninae), which often have horns on their head and thorax. Many fungivorous beetles are found on their host fungi all year round in temperate countries, and some such as Tetratomidae and Phloiophilidae are found most easily in winter.

Specialized fungivorous beetles feed and develop on fresh fungi, but when the fungal fruiting bodies start to decay, fly maggots and a wide range of predatory beetles and scavenging beetles arrive, including Staphylinidae, Silphidae, and Histeridae. Other families such as Latridiidae and Ptiliidae may feed on fungal spores.

Smaller fungi support a range of different beetle families: sooty bark disease, a fungus that causes a black mold under the bark of maples, particularly sycamore maple, has a very characteristic suite of beetles of the families Zopheridae and Latridiidae associated with it, which are rarely, if ever, seen anywhere else. Many beetles actively spread fungi from tree to tree, in some cases acting as vectors that can spread a pathogenic fungus from an infected tree to a healthy one. A famous example are the elm bark beetles (genus *Scolytus*), which have mycangia, specialized pockets that carry fungal spores with which they inoculate their burrows to grow a fungal crop for their larvae to feed on. This has resulted in the spread of Dutch elm disease. Recent DNA studies have shown that most wood-feeding beetles, even those lacking mycangia, carry on their bodies the spores of many species of wood-decay fungi. This may

be a strategy by the fungi to use the beetles to transmit their spores, in a more directed way than releasing them into the wind, since the beetle and the fungus both require the same resource. It may be a complex symbiosis, where the fungi may make the wood more digestible for the beetle larvae, while the beetles provide access and break up the wood, increasing its internal surface area with their tunneling.

Some fungi have turned the tables on beetles and become parasites. Fungi of the order Laboulbeniales resemble bristles or scales that grow on the exoskeleton. They are transmitted from adult to adult, so they are generally found only in beetle families with long-lived adults that routinely contact adults of the next generation. These include Carabidae, Dytiscidae, and Coccinellidae. Ladybugs often have small patches of these fungi on their brightly colored elytra.

DEAD ANIMAL TISSUE

Beetles are often called "recyclers" or "nature's cleanup crew" because so many of them play a role in consuming or generally clearing up dung, dead plants, and animals. However, the whole concept of "ecosystem services" is somewhat anthropocentric. The beetles are not serving a preordained useful function but are simply taking advantage of a nutritious resource. The fact that we should be grateful for this, and may actually depend on this recycling of nutrients for our continued existence, is purely fortuitous.

Of the dead materials that different groups of beetles feed on, decaying animal matter is one of the most digestible and, like dung, is thinly distributed in the environment. Beetles that specialize on carrion need powerful senses to detect it and good mobility in order to get to it quickly. Usually they are attracted by the smell of decay, which means that bacteria and fungi are already established on the carrion. Some carrion beetles, for example Silphidae, treat the carrion on arrival with antibacterial and antifungal secretions to discourage microbial growth and keep it fresh for use by their own larvae. Many carrion beetles also carry on their bodies phoretic (that is, hitchhiking, nonparasitic) mites, which are flightless and use the beetles as a vehicle to get to the carrion. They then benefit the beetles by

BELOW | *Dermestes maculatus* (Dermestidae) Larvae of this hide beetle are used to clean the flesh from a vole skull for a vertebrate collection at University of California, Berkeley.

eating the eggs and small larvae of flies and other insects that might otherwise compete with the beetles. The Silphidae genus *Nicrophorus*, the burying beetles, will proceed to bury the carrion if it is small enough, to get it out of the reach of vertebrate scavengers. They then skin it, lay eggs, and prepare the buried carcass for their larvae. Silphidae and several large, carrion-feeding Staphylininae (Staphylinidae) also produce chemicals that are deliberately repellent to mammalian scavengers, to try to prevent the carrion, together with the silphid eggs and larvae, being eaten by a larger scavenger.

Carrion at different stages of decay is used by different groups of beetles. For example, fresh or slightly decayed flesh may be attractive to Silphidae, Geotrupidae, Histeridae, and Staphylinidae, which may feed on either the carrion itself or on maggots and other invertebrates that are attracted to it, but there is a whole spectrum of carrion feeders until the late stages of decay, when only skin, hair, and bones remain. These later stages may attract Nitidulidae, Dermestidae, and Trogidae, which are basically keratin feeders, and are also found on carnivore dung filled with undigested hair, or debris in bird nests consisting of feathers. Colonies of Dermestidae are kept by museums to prepare skeletons, as the beetle larvae strip off all remains of skin and dried muscle, leaving only the clean bones behind.

In South America, with the extinction of the ground sloths and other megafauna, many dung beetle groups have adapted to feeding on carrion. For example, the giant metallic horned dung beetles *Coprophanaeus ensifer* and *C. lancifer* (Scarabaeidae: Scarabaeinae) are almost entirely necrophagous.

VERTEBRATE DUNG

Being vertebrates ourselves, we might consider vertebrate dung to be a worthless waste product, but to many insects with finer digestive systems, this mixture of incompletely digested food, bacteria, and fungi is a valuable resource, and hundreds of insect species have adapted to exploit it. Chief among these are beetles.

Dung can be rather scarce and widely scattered in the ecosystem, so a dung-feeding insect needs well-developed senses to detect it quickly, and good mobility to get to it while it is still available. Scarab dung beetles (Scarabaeidae: Scarabaeinae) use the scent-detecting flaps on their lamellate antennae to sense dung over long distances, and fly powerfully toward it. Dung beetles are so well adapted that they can start arriving at a dung pile within moments of it hitting the ground, and immediately start to compete among one another for this precious resource, which they use for breeding as well as eating it themselves. One strategy, used by so-called "rollers," is for a female or a pair (depending on the species) to form as much dung as possible into a ball, often larger than the beetles themselves, and roll it away from the melee to a safe place, where it is buried and a single egg laid inside. The dung beetle larva then develops in safety surrounded by food in the ball of buried dung, while the adults return to the pile to repeat the process.

A single large ruminant such as a cow or an antelope can produce 44–66 lb (20–30 kg) of dung per day, and so the great herds, for example of Africa, can support millions of individual dung beetles of hundreds of species. In healthy, balanced ecosystems, deposits of dung even from large animals such as elephants may be cleared up within minutes, refertilizing the soil in the process

BELOW | *Emus hirtus* (Staphylinidae) Not all dung-associated beetles are scarabs. This is one of the largest rove beetles and a bumblebee mimic—it is a predator of fly maggots.

and supporting the grasses on which the animals depend for food.

When Europeans brought animals such as sheep and cattle to Australia, the dung ecosystem there was unbalanced, because although the settlers accidentally brought some species of flies, they did not think to bring any dung beetles. It was soon found that the Australian dung beetles, adapted to the smaller, drier droppings of marsupials, were not interested in the huge deposits made by the larger imported animals, and so the cattle and sheep dung lay and dried on the pastures, covering the grass and breeding flies. In the end, the only way to make cattle and sheep agriculture sustainable in Australia was for the government to employ entomologists to select, breed, and release masses of dung beetles that were specialized for the dung of the imported ruminants. This became a famous example of successful biological control, and saved the Australian cattle and sheep industries. It also showed the importance of insects in ecology, which may not always be recognized until they are not there.

ABOVE | *Pachylomera femoralis* (Scarabaeidae) A mass of beetles compete for a dung pile. They start arriving while it is still warm, and may clear it away before it has cooled.

BELOW | *Canthon quadriguttatus* (Scarabaeidae) In the rainforests of Peru, some adult dung beetles cling to the undersides of brown titi monkeys, so they can beat the competition when substrate emerges.

BEETLES AS PART OF THE FOOD WEB

With hundreds of thousands of species, and billions of individuals occupying a huge range of ecological niches, it is not surprising that beetles form an essential part of most terrestrial food webs. For example, a recent study from a Swiss university estimated that birds around the world eat 440 to 550 million tons (400 to 500 million tonnes) of beetles and other arthropods per year. It is difficult to imagine the number of individual insects in 550 million tons, but it certainly puts the collecting of a few thousand specimens by entomologists into perspective!

If you were to remove beetles, many terrestrial ecosystems would rapidly collapse. A great deal of the concern in the media about recent evidence of insect decline is not concern

for insects as such but rather a fear of the knock-on effects of insect loss on more visible and "popular" organisms, such as the birds and mammals that eat them, or the trees and shrubs that they pollinate. However, a strong decline of insects—even just beetles—would have far-reaching implications that would impact the overall function of global ecosystems.

As well as providing food for many other invertebrates and vertebrates, and playing little recognized but major roles as pollinators and predators, beetles and their larvae have an essential role to play in returning nutrients to the soil and making them available again for new plant growth. European settlers in Australia received just a small taste of what could happen if these systems are unbalanced, when they introduced sheep and cattle to Australia without

the suitable dung beetles to clear up the waste. What was almost an ecological disaster was averted by the selective introduction of European and African dung beetles, but if this had not been done, or had been unsuccessful, the entire Australian sheep and cattle industry might ultimately have become unsustainable. This imbalance could be fixed only because it was on a continental and not a global scale, but if dung beetles were to disappear globally, so that more could not be brought in from somewhere else, humanity could be faced with an agricultural crisis. This scenario may not be as distant or unrealistic as it sounds. For example, the widespread use of ivermectin wormers to kill intestinal worms in livestock may be destructive to dung beetle populations because these pesticides are nonselective and remain active in the dung after it has passed through the animal, killing many of the insects that would otherwise break down the dung.

The famous US entomologist E.O. Wilson (1929–2021) described insects as "the little things that run the world," and this quotation is very apt. When we look at a tropical rainforest, we see trees and flowers, and we hope to see, perhaps, orangutans, tapirs, or jaguars. But what is much more fundamental is the vast, underappreciated, and in many cases still undescribed diversity of beetles that, in their millions, are maintaining the whole ecosystem.

OPPOSITE ABOVE | *Dicerca* (Buprestidae) A hunting wasp captures a jewel beetle. This specialist wasp only hunts jewel beetles, and so also helps scientists monitor the spread of the invasive Emerald Ash Borer *Agrilus planipennis*.

LEFT | A Namaqua Chameleon catches a toktokkie desert beetle (Tenebrionidae) in the Namib desert.

SPECIALIZED HABITATS

VERTEBRATE NESTS

The nests of vertebrates, here not including human dwellings, which are discussed separately (see page 60), provide multiple opportunities for adults and larvae of beetles. Nests are built by animals to live in and rear their young. As well as the bedding material used to make the nest itself, the nests usually contain numerous potential food sources for beetles. These include leftover or discarded food, waste and excrement, feathers or fur, fungi and molds, the occasional unhatched egg or dead nestling, and a community of other invertebrates as scavengers or parasites. The nest will usually have high humidity, and if the animal is warm-blooded, such as a mammal or bird, a warm microclimate. This may even be the case with incubation nests of cold-blooded animals such as crocodiles, snakes, or turtles that lay their eggs in mounds of decaying vegetation, or exposed to the sun. The presence of a permanent resident mammal or bird also provides some protection, especially if the resident animal is not an insect eater.

A rich and varied beetle fauna can usually be found, alongside fleas, flies, their maggots and other insects, in a typical mammal or bird nest. Members of some families such as the Leiodidae, Histeridae, Trogidae, and some subfamilies of Staphylinidae and Scarabaeidae, are strongly specialized for the nests or burrows of particular animal species, and they are seldom found anywhere else. For example, a striking and previously unknown diversity of small dung beetles of the scarab subfamily Aphodiinae has recently been discovered in the burrows of pocket gophers in supposedly well-known and well-studied parts of North America, and none of the new species had previously been discovered in the dung of surface-dwelling rodents or other animals. Examination of a bird nest after the chicks have

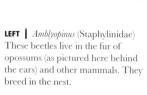

fledged can also reveal large numbers of small
beetles and their larvae feeding on the debris left
behind, or as predators on other nest inhabitants.
The majority of beetles in vertebrate nests are
scavengers, and are harmless or even useful (for
example, by clearing up dung or feeding on
parasites or fungi) to the nest builders.

There are a few species, however, that have
crossed the line into harming their hosts. Burying
beetles of the genus *Nicrophorus* (Silphidae), usually
scavengers, can be attracted to vertebrate nests to
feed and rear their larvae on young that have died
or unhatched rotting eggs. The North American
species *Nicrophorus pustulatus* has been found to
attack live, healthy snake eggs within the snakes'
nesting mound, the beetle larvae developing inside
the eggs and killing them in the process, acting
as a parasitoid instead of a scavenger. Some
Trogidae such as *Omorgus suberosus* may do the

same thing in the nests of sea turtles, and may
pose a risk to some threatened species.

Only very few genera of Staphylinidae and
Leiodidae have made the step to living on the host
animal itself, and even these are not usually true
parasites; instead they feed on flakes of skin, hair,
and other organic matter, or even on other insects.

ANT NESTS AND COLONIES

Modern social ants arose from wasp-like ancestors about 100 million years ago during the Cretaceous period, when dinosaurs walked the earth. They rose to ecological dominance, and still play a major role as predators and scavengers in most land ecosystems. An ant colony is a superorganism controlled by a reproductive queen which produces thousands of infertile workers that forage and hunt for the colony, defend it, and rear more workers. There is a view that an ant colony is a single composite organism, because the queen is the only individual capable of reproducing, and the workers produced and controlled by the queen are like an extension of her body. Based on the combined weight of their workers or the volume of food they consume, ant colonies are some of the largest predators on earth. Troops of workers hunt other invertebrates and sometimes even vertebrates: African driver ants can overcome big animals such as dogs and goats that are caged or tethered, and their marching columns are used by African people to clear their homes or crop fields of pests. These driver ants, and their South American equivalents, the army ants, have no fixed nest, but carry the queen and larvae with them in ever-marching columns. Most other ants have a vigorously defended nest, to which they

chemical and physical—to persuade ants to ignore them, mistake them for food, or even welcome them as nest mates, as well as other adaptations such as shortening of limbs and antennae, and fusing of abdomen segments, to protect them from damage during rough handling by the ants. Animals that have adapted to live with ants are called myrmecophiles (ant lovers) and can be recognized by characteristics including clumps of setae called trichomes that provide food or chemicals to the ants, fused segments, short antennae, sometimes loss of pigmentation or eyes, and, in some extreme cases, structures that allow the ants to pick them up and carry them without damage. We can even see such characters in fossils, and can assume they must be myrmecophiles without knowing their actual biology.

Hundreds of species of beetles of many families have become myrmecophiles, the most common of these being Histeridae, Carabidae: Paussinae, some Scarabaeoidea, and particularly Staphylinidae of the subfamilies Aleocharinae and Pselaphinae. The Pselaphinae tribe Clavigerini are almost all compulsory ant associates. Some myrmecophiles are scavengers or cleaners, while others consume the ants' resources, and a few eat the ants themselves as well as their larvae.

OPPOSITE | *Paussus* (Carabidae: Paussinae) This highly modified ground beetle, almost unrecognizable compared to its surface-living relatives, has many chemical and physical adaptations in order to survive in ant nests.

ABOVE | *Limulodes* (Ptiliidae) Featherwinged beetles, these from the USA, act as cleaners in ant nests, eating fungi and spores.

bring their prey. Not all are predators: the South American leaf-cutter ants form huge colonies with subterranean fungus gardens, where they bring slices of leaves to rot and grow the fungi that they consume. They can defoliate whole trees.

Ant colonies represent a great opportunity: safe, climate controlled, full of stored food and ant larvae, but any beetle taking advantage of them needs to get past the ants. Beetles of several families have developed a suite of adaptations—

TERMITE NESTS AND COLONIES

Tall edifices, seemingly built of stone, are characteristic features of the dry grasslands of the African savanna, the Brazilian Cerrado, and the Australian outback. These are the colonies of termites, each one built and maintained over decades or centuries, and home to thousands or even millions of termite workers. The structures cover extensive subterranean networks of chambers and tunnels. The mounds are built from soil cemented with the saliva and dung of the termites, and are surprisingly strong and weatherproof, although they need regular maintenance by the workers.

Termites are social insects, living in colonies controlled by a reproductive pair, called the "queen" and the "king." The white, sausage-like queens grow huge and are some of the longest lived of insects, typically exceeding ten years and reported at up to fifty. When the queen dies, the supply of a hormone she produces inhibiting reproductive development in the colony stops, and another queen arises and produces the same hormone to inhibit the rest of the colony. Not all termite genera inhabit giant fortified nests; many

live in the soil or in arboreal nests made from a lightweight substance called "carton."

Termite colonies have been called the world's oldest societies, with the earliest fossil termites known from the beginning of the Cretaceous period. Evidence suggests that the group is even older; one piece of evidence for the early appearance of termites is a Late Jurassic fossil mammal *Fruitafossor windscheffeli*, described in 2005 in North America, which has been interpreted as a specialist termite predator from its tubular teeth and digging limbs, resembling modern termite eaters.

Termites were for a long time placed in their own order of insects, Isoptera, but recent research combining DNA, fossil, and morphological evidence has shown that they are eusocial relatives of cockroaches (Blattodea). Despite similarities in nest and social structure, they are not closely related to ants. However, like ants they are divided into castes, with the majority of the colony being nonreproductive workers, which maintain the structure of the colony and collect food, or soldiers, which defend it, using large jaws or chemical defenses. The similarity has led to termites being called "white ants" in many parts of the world.

Also like ants, the colonies of termites provide abundant opportunities for nest invaders (called inquilines) to occupy their living space, taking advantage of the constant temperature and humidity, protection from predators, or the reliable supply of food (whether the food of the termites or the termites themselves). Many beetles

LEFT | *Termitotrox cupido* (Scarabaeidae) This blind, flightless scarab beetle was discovered in 2012 in termite nests in Cambodia and named after a Roman god of love.

Penichrolucanus copricephalus
(Lucanidae) This very small and
uncommon hornless stag beetle from
Southeast Asia is associated with
termite nests.

have adapted to circumvent the
termites' defenses. Some have
only a superficial relationship,
being scavengers in and around
the nest, while others have
complex physical and chemical
systems to deceive their hosts,
and are fully integrated into the
colony. An extreme example are
the larvae of the Australian
Megaxenus (Aderidae), which imitate
the queen termite so that the workers
bring them food and clean them. Click beetle
larvae of the genus *Pyrearinus* (Elateridae) live
in burrows on termite mounds in Brazilian
grasslands, using bioluminescence to attract
prey to their powerful jaws.

BELOW | *Pyrearinus termitilluminans* (Elateridae)
Bioluminescent click beetle larvae light up a termite
mound in Brazil's Pantanal. They are predators that
use light to attract flying insects, including termites.

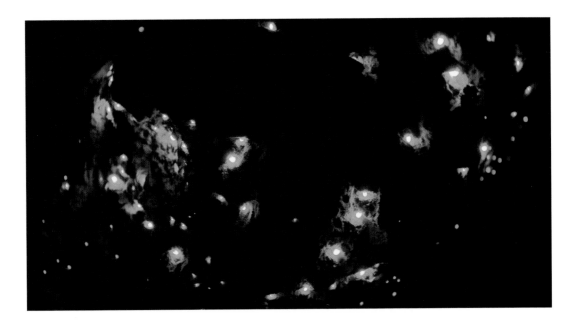

CAVES

A whole field of biology called biospeleology is dedicated to the exploration, discovery, and study of organisms that inhabit caves. Caves provide a very stable environment: in their deepest parts almost totally isolated from surface influences; but the cold, damp, and perpetual darkness, and limited supply of nutrients, make them hostile for most life. There is a spectrum of cave insects, from surface fauna that has temporarily moved into the shallower and more accessible parts of caves, to so-called true troglophiles, species that have adapted over long periods of time to cave conditions. The latter, among beetles, are either predators (such as the families Staphylinidae and Carabidae) or scavengers (Leiodidae), since the ecosystems lack living plants (which need sunlight for photosynthesis). Instead of plants, cave ecosystems are based on detritus and organic matter either carried in by water or deposited by larger inhabitants, such as the dung and carcasses of bats or cave-dwelling birds. True troglophiles have a distinctive set of adaptations: because of the darkness, they often lack eyes and pigmentation. To compensate for the lack of light and the uselessness of eyesight, many have developed long, slender limbs and antennae to enable them to feel for food.

Experts studying cave fauna often say that those large caves that humans can enter and explore are, from an insect's point of view, just the most accessible of many holes in a complex network of interconnected crevices, gaps, and fissures in the porous rocks. The troglophile

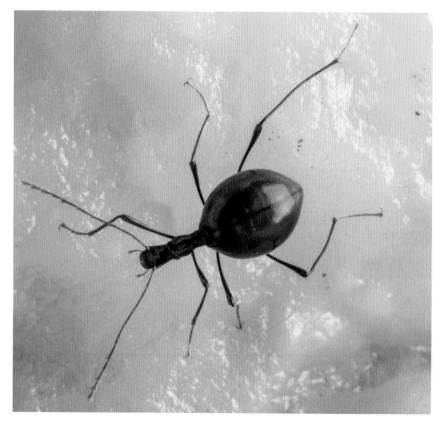

LEFT | *Leptodirus hochenwartii* (Leiodidae) A true troglophile, this eyeless beetle from the Dinarides in the Balkans shows very long limbs and antennae.

OPPOSITE | *Duvalius gebhardti* (Carabidae) An eyeless ground beetle from caves in Hungary. Many troglophiles lose eyes and most of their pigmentation.

cave fauna, then, are just a part of a sometimes inaccessible interstitial fauna that exists over a much wider subterranean area. If this is the case, then the fauna of a particular cave may not be as isolated as it initially seems, since it may be connected with neighboring caves by cracks and tiny corridors that allow insect populations to pass from cave to cave without having to go outside.

In northern parts of the northern hemisphere, such as Britain, caves are common in chalk and limestone districts and are well studied, but no associated troglophile cave fauna are known. This is thought to be because they were scoured and sterilized by the cold and ice of recent glaciers, and the cave insects have not had any opportunity to recolonize. It is interesting to speculate what might have inhabited these caves before the glaciations of the last Ice Age. Farther south in

Europe, for example in Hungary, the Balkans, Italy, or Spain, and at equivalent latitudes in Asia and North America, a rich troglophile fauna of detritivores and small predators adapted to the environment are found, and many unusual new species are being discovered.

The low level of available nutrients means that true cave fauna usually has low populations which are widely spread within the cave and surrounding soil. Biospeleologists use traps set over many months—baited with strong-smelling foods and drinks such as cheese, beer, and wine—in order to lure individuals of this fauna, but even after a year or more a trap may yield only a small number of specimens.

SAND

Loose sand is another specialized habitat that many organisms have failed to colonize, but which supports a varied range of beetle species. Two major environments consist largely of sand: beaches, which may merge into sand dunes, and deserts. While these two habitats have certain features in common, there are many differences, and they are used by different groups of beetles.

Sand, whether on deserts or beaches, contains little or no organic matter, so plants do not grow well on it. As a result, there is little shade, and inhabitants are exposed to the sun and to predators during the day, unless they can hide beneath debris or bury themselves. Sand also does not retain moisture well, so such habitats are usually very dry. Therefore, beetles that inhabit any sandy environments, especially desert beetles,

need to be drought tolerant. There is also the question of where food comes from, considering the lack of vegetation.

Beetles that inhabit beaches often obtain their nutrition indirectly from the sea, feeding on marine life that has been washed up on the beach. Whether piles of seaweed or the carcasses of marine animals, ranging from jellyfish to whales, most decaying organic matter is attractive to some beetle scavengers, and often to predatory beetles that eat the larvae of other insects, particularly flies, which can be hugely abundant. The sea may also deposit land vegetation, fruits, drowned land animals, and driftwood, all of which are exploited as a food source by some beach-living beetles. Some beetle species specialize in driftwood, and have been transported around the world in floating timber.

The majority of beach-living beetles belong to the families Carabidae, Staphylinidae, Histeridae, and Tenebrionidae, though some smaller families are also characteristic of beach habitats, for example the Phycosecidae in Australia. The inhabitants of driftwood include wood-boring Curculionidae and Oedemeridae. Many carabids and staphylinids on sandy beaches also feed directly on marine organisms such as sandhoppers (Crustacea: Amphipoda). Almost all

LEFT | *Stenocara eburnea* (Tenebrionidae) One of the white, desert-adapted Pimeliinae, this beetle is from Africa's Namib desert.

beach-living beetles obtain their moisture from their food.

In deserts, there are much fewer options as there is no sea to supply organic matter. Almost all desert beetles in true deserts belong to the family Tenebrionidae, most frequently the subfamily Pimeliinae. They are flightless, with fused elytra, and are covered with a waxy substance to prevent evaporation. In Africa, species of the genera *Stenocara* and *Onymacris* bury themselves in sand to avoid the heat of the sun, and have long legs to raise their bodies off the hot sand surface. They scavenge the desert for fragments of plant and insect matter that have blown in, or in the case of insects, flown in and died. Of course, such material is completely dry, so many species of Tenebrionidae in the world's driest deserts need to obtain liquid from the atmosphere, which they do by the unique behavior of "fog basking," harvesting minute droplets of water from the atmosphere at specific times of day.

FLOWING FRESH WATER

Like most non-marine habitats, flowing fresh water such as streams and rivers is home to many beetles, but fewer than are found in ponds, ditches, and slower-moving water. There are several ecological obstacles that keep some beetle genera and species out of flowing streams. Firstly, most larvae and adult water beetles need to return to the surface fairly often to replenish their supply of air, and in flowing water this exposes them to the risk of being carried away by the current. Secondly, the quantity of organic matter, and so the potential food supply, is usually less in faster-flowing water. Thirdly, fast-flowing water has higher levels of oxygen and is able to support larger populations of fish, many of which will eat the adult and larval beetles. Despite these hurdles, a number of families and genera of beetles have adapted to survive in these environments.

The family Amphizoidae are so closely associated with fast-flowing streams that they are

called trout stream beetles. They are large water beetles, 10–15 mm long, represented in the modern fauna only by the genus *Amphizoa*, with five species: three in North America and two in east Asia. They inhabit clean, well-oxygenated, rapid-running mountain streams, where adults and larvae are predators of immature stages of insects such as caddisflies and stoneflies. The long-lived larvae of water penny beetles (Psephenidae) cling to the undersides of rocks in flowing streams, grazing on algae, and are found in small numbers throughout the world. They are so called because their circular larvae resemble small coins. The small, soft-bodied adults are not aquatic, and live short lives on streamside vegetation.

Another beetle group that received its common name from its association with flowing water is the Elmidae, called riffle beetles. Adults and larvae live under stones in fast-flowing streams, grazing on algae and other encrusting organisms. A few elmid genera, such as the usually rare *Stenelmis*, develop as larvae in submerged dead wood at the bottom of streams and rivers. Several genera of Elmidae have developed a "plastron," which enables them to extract oxygen direct from water, removing the need to go to the surface to recharge their air bubble. Elmidae are much more diverse in fast-moving water than they are in ponds or pools, which are preferred by other water beetle families such as Hydrophilidae, Dytiscidae, and Gyrinidae. However, all of those families also have a few species and genera that specialize in moving water.

LEFT | A typical European riffle beetle *Limnius volckmari* (Elmidae), grazing algae on a submerged rock.

The banks of flowing waters, where there is mud periodically washed by the stream, support large populations of many beetle families. Dryopidae, Heteroceridae, Hydrophilidae, Limnichidae, Hydraenidae, Staphylinidae, and Carabidae can be abundant in temperate stream banks, and can be collected by standing in the stream and splashing water onto the mud, which brings them out of their burrows. In the tropics, larger carabids such as the genera *Scarites* and *Galerita* and tiger beetles (Cicindelinae) can be found in such situations.

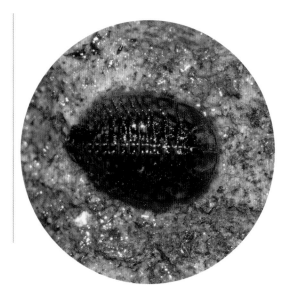

SALT WATER

The open sea is one of the last great habitats that has never been conquered by beetles, although beaches, sand dunes, and even the strandline support a rich and varied beetle fauna. Many beetles live in fresh water, but these are descended from land-living ancestors, and the larvae of almost all freshwater beetles have to leave the water to pupate on land, even those which then return to the water as adults. Pupation on land is probably too difficult and unreliable in marine environments, where the distance to the nearest dry land may be enormous and unpredictable.

However, since insects in general are almost absent from the sea, even those insects that do not have a pupal stage, the pupation site cannot be the only reason for the lack of truly marine beetles. There must be other factors that keep them out. It is likely that, having evolved and diversified on land, insects lack physiological adaptations for life in salt water; also most of the available ecological niches are already occupied by crustaceans, which were there first. One of the only truly offshore oceanic insects, not beetles but true bugs, the ocean strider genus *Halobates*

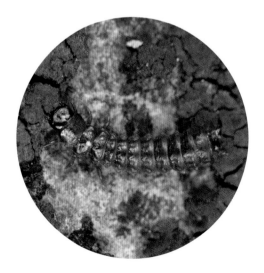

ABOVE | *Cicindis horni* (Carabidae) A rare semiaquatic beetle of the Salinas Grandes salt pans of Argentina, which enters the salt water to hunt for fairy shrimps.

(Hemiptera), skate on the water's surface, so they are really inhabitants of the air rather than the ocean, which is beneath their feet.

Although none have colonized the open sea, numerous families of beetles have adapted to and specialized for life on the shoreline, for example huge numbers of Anthicidae, Staphylinidae, Ptiliidae, and Histeridae can be found in piles of rotting seaweed cast up above the high-tide mark, where they feed on fly larvae or decaying organic matter. These beetles can survive temporary immersion, and if dropped into seawater usually climb onto floating debris or skate on the surface tension and immediately take flight.

Other groups, particularly Carabidae, are nocturnal hunters; they shelter under beached flotsam during the day, and patrol along the strandline at night for sand hoppers and other crustaceans or marine worms that have ventured up onto the beach. These include *Eurynebria complanata*, a large and striking, pale yellow and black inhabitant of Atlantic and Mediterranean

coastlines of Europe and North Africa. *Eurynebria* has disappeared from much of its former range in recent years, and the reasons for its decline are not fully understood, as in some cases the beaches it inhabited seem unchanged to the human eye. A few other carabid ground beetles take their association with water even further, submerging themselves temporarily to pursue their aquatic prey in its own environment.

Some water beetles of several families have adapted to live in brackish or even salt water, but these generally only inhabit tide pools in the splash zone on rocky shores, which are rarely inundated by the waves, and take shelter in cracks in the rocks during exceptionally high tides. Other beetles, such as some small Carabidae, Staphylinidae, Melyridae, and Salpingidae, live permanently on rocky shores between the tide lines, emerging at low tide to feed on algae or stranded marine organisms, and returning during high tides to fissures in the rocks where some air is trapped.

TOP | *Ochthebius marinus* (Hydraenidae) Warm, sun-exposed tide pools that are only occasionally refreshed by the sea provide ideal habitats for the European marine moss beetle.

ABOVE | *Ochthebius marinus* (Hydraenidae) A small water beetle adapted to pools of brackish or salt water near the sea, where it grazes on algae.

OPPOSITE BELOW | *Aegialites* (Salpingidae) The larva of a genus of small, flightless beetles. The adults and larvae live in cracks in intertidal rocks on both sides of the North Pacific.

BEETLES AND HUMANS

HISTORY OF COLEOPTEROLOGY

The history of the study of Coleoptera dates at least to the Classical period of ancient Greece and Rome. Aristotle (384–322 BCE) and his teacher Plato both mentioned beetles, and many of the generic names used by Carolus Linnaeus, such as *Buprestis* and *Cicindela*, are from *Naturalis Historia* by Pliny the Elder (23–79 CE), the largest surviving book from the Roman Empire. After the fall of Rome, Europe suffered an extended period of comparatively little curiosity-driven academic study, and during these Dark Ages, interest in beetles was restricted to their uses or threats to human health or food security, or to allegorical or superstitious significance. During the Medieval period, monastic texts repeated classical authors, often with stylized images, and ascribed spurious medical properties or even religious significance to beetles.

A growth of learning and curiosity about nature marked the Renaissance and the Enlightenment, and by the late 1600s it became more usual for educated, affluent people to have cabinets of curiosities, the progenitors of museum collections. In 1735 Linnaeus published *Systema Naturae*, which provided a framework and system

ABOVE | *Chiasognathus grantii* (Lucanidae) Darwin's Stag Beetle. Charles Darwin, during HMS *Beagle*'s stop in Chile, was among the first to observe the behavior of this beetle.

LEFT | *Batocera wallacei* (Cerambycidae: Lamiinae) From New Guinea, this is one of the largest of its genus, named after Alfred Russel Wallace.

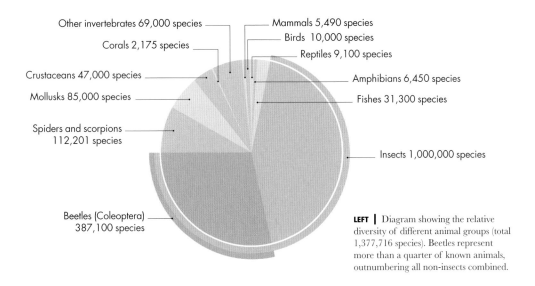

Other invertebrates 69,000 species

Corals 2,175 species

Crustaceans 47,000 species

Mollusks 85,000 species

Spiders and scorpions
112,201 species

Beetles (Coleoptera)
387,100 species

Mammals 5,490 species

Birds 10,000 species

Reptiles 9,100 species

Amphibians 6,450 species

Fishes 31,300 species

Insects 1,000,000 species

LEFT | Diagram showing the relative diversity of different animal groups (total 1,377,716 species). Beetles represent more than a quarter of known animals, outnumbering all non-insects combined.

to classify living organisms, and he accumulated, and in 1758 named, several hundred beetle species. This led to a flowering of interest, as the new naming system was applied ever wider, and sailors and explorers brought back natural objects from far away. Captain James Cook (1728–79) included a naturalist, Joseph Banks (1743–1820), on his famous *Endeavour* voyage, ultimately to New South Wales, Australia. Most of Banks's collections survive, as one of the oldest modern-style scientific collections in the Natural History Museum in London, UK.

Specimens collected by Pierre Dejean (1780–1845), an important entomologist and officer in Napoleon's army, still exist. He supposedly got off his horse during the Peninsular War Battle of Alcañiz to collect a Cebrioninae (Elateridae), which he pinned into his helmet. After his defeat, he was pleased to observe the beetle was still in good condition, and it is now in a museum in Torino, Italy.

By the nineteenth century, the study of beetles had become widespread and popular both as a

formal discipline and as a scientific pastime, where affluent private collectors assembled important collections. Countries made collections too, building natural history museums to house the accumulated specimens, generate knowledge, and educate and inspire the population. This environment produced Charles Darwin and Alfred Russel Wallace, two young men who shared a burning interest in collecting beetles and went on to change the world. Darwin said in his autobiography, about his teenage beetle-collecting days, "It seems therefore that a taste for collecting beetles is some indication of future success in life!" He and Wallace are good examples of such success, starting out as beetle collectors and raised to greater heights than anyone could imagine from a platform of dead beetles.

The study of Coleoptera goes on, and museum beetle collections provide vast archives of the beetle knowledge accumulated across the world and down the centuries, ready to answer more questions, some of which society has not yet even thought to ask.

PESTS OF CROPS

One of the reasons for the great success of beetles is the close association of many families with plants, so that almost every plant genus has several species of beetles feeding on it, many of which are host-specific. Furthermore, different beetles will utilize different parts of the plant, which means that a given host plant may support a range of beetles. This intense herbivory is, of course, damaging. Many plants have attempted to avoid it by becoming annual, growing from a seed to a mature plant that produces its own seeds in a single season, then dying off, with its offspring growing some distance away next season. Beetles have responded to this by becoming more mobile and developing senses that allow them to "smell" their target host plants and home in on them over long distances.

During the Neolithic Revolution, some 12,000 years ago, when people started to live in settled societies, they domesticated the first few plants as food crops, and many of these were annual cereals or legumes with large seeds. Arable farming can be an invitation to pests, because farmers grow large numbers of plants of a single species close together in monocultures, in the same place every year. It seems likely that crop pests have been with us as long as agriculture has, and this is supported by evidence of peas bored by Chrysomelidae: Bruchinae (probably the species *Bruchus pisorum*) from archeological sites in Jordan and Turkey some 8,000 to 9,000 years old.

As the selection of crops we grow has increased, so has the selection of potential pests, especially when crops are transported to new areas. A famous example is the Colorado Potato Beetle *Leptinotarsa decemlineata* (Chrysomelidae), a leaf beetle native to the Rocky Mountains of North America, where its original host plant was a native weed, buffalo bur. During the nineteenth century, with widespread planting of potatoes imported from the Andes, the beetle switched host to this new, abundant resource, and within 20 years it was found on potatoes from coast to coast of continental USA. In the 1920s it reached Europe and spread east, until it occupied a belt around the whole northern hemisphere,

LEFT | *Leptinotarsa decemlineata* (Chrysomelidae) The strikingly marked Colorado Potato Beetle is a scourge of potato agriculture across the northern hemisphere.

covering an area of over 6 million sq miles (16 million sq km). It prefers continental steppe climates with cold winters and hot summers, avoiding the warm, damp climate of Britain and Ireland. There was an unsuccessful plan by Germany during the Second World War to introduce Colorado Potato Beetle to England, to destabilize enemy food production. Similarly, when it arrived in East Germany during the 1950s, propagandists tried to blame the new pest, groundlessly, on American interference, calling it "Amikäfer" (Yankee beetle).

The Cotton Boll Weevil *Anthonomus grandis* (Curculionidae) is a serious cotton pest in southern USA, recently doing up to $300 million worth of damage per year. Strangely, there is a monument to this pest in the town of Enterprise, Alabama, because its depredations forced the area to diversify into other crops, bringing prosperity.

PESTS OF STORED AGRICULTURAL PRODUCTS

With the agricultural revolution, or even earlier with hunter-gatherers, people stored food from times of plenty for future times of need. Like almost every activity, this practice provided opportunities for beetles to exploit. When people travel from place to place, especially large groups over great distances, they usually bring a supply of preserved food for the trip. If food is carried for periods of time in open conditions, or by ship, it might get damp or spoiled, and on reaching the destination will be discarded, distributing any insects it contains into a new environment. As humans have spread all over the world, the insect travelers associated with their food supplies have been carried with them, and at each destination, more species were picked up accidentally and added to the assemblage of "human stored product beetles."

In this way, numerous beetle species that feed on preserved or dried grain, fruit, fish, or meat have been spread throughout the populated world. Remains of the exact same pest species have been found preserved in the pyramids of ancient Egypt and in Native American archeological sites, showing how long they have been with us and how far we have traveled together. The original ranges of these beetles are lost in time, but we can infer their original habitats, since humans are not the first animals to store products, and the habitats provided by human food storage—effectively

dark, dry places full of a single kind of food—are not unique in the natural world. Some birds such as jays hoard seeds in hollow trees; rodents such as squirrels, mice, rats, and hamsters store grain in their burrows, and it is thought that some of the most serious pests of grain storage were originally associated with the underground larders of small mammals.

To be associated with stored products, beetles need three main characteristics: to be resistant to dry conditions as adults and larvae, since stored products are, where possible, kept dry; have the ability to disperse over large distances, as supplies of food are scattered in the environment; and for the same reason, like any animal needing an infrequent resource, be able to detect reliably the resource that they will infest. Families of beetles that meet these criteria, and members of which have become cosmopolitan stored product pests, include Tenebrionidae, Laemophloeidae,

RIGHT | *Oryzaephilus surinamensis* (Silvanidae) The earliest known record of the cosmopolitan Saw-Toothed Grain Beetle is from a Neolithic site in northern Greece.

Silvanidae, Latridiidae, Cryptophagidae, Nitidulidae, Bostrichidae, Ptinidae, Chrysomelidae subfamily Bruchinae, and Curculionidae.

The combined costs of these pests, not just financial but in terms of lost livelihoods and even lives, is enormous, especially in tropical countries, where a hot, humid climate promotes insect and fungus growth, and the infrastructure may make safe storage more difficult. The Western Australian Department of Agriculture and Food estimates that a quarter to a third of the world's entire grain crop is lost in storage. The United States Department of Agriculture claims that 10 percent of all food produced in the USA is lost to pests or pest contamination, and that this may rise to over 50 percent in some countries. Although other insects and fungi play a role, beetles are major culprits in the destruction of stored products, costing millions of dollars in damage every year.

ABOVE | *Tribolium castaneum* (Tenebrionidae) The Red Flour Beetle is a major pest of starchy foodstuffs, and has been spread throughout the world.

BELOW | *Callosobruchus maculatus* (Chrysomelidae) A bean beetle infesting mung beans. All they leave behind is a husk packed with droppings that even chickens won't eat.

FORESTRY PESTS

A natural forest is a rich mosaic of habitats, where trees of many species and varying ages, from seedlings to ancient veterans, live with a diverse assemblage of other plants, fungi, mammals, and birds, and hundreds or thousands of species of invertebrates, forming a complex and balanced ecosystem. A plantation forest grown for forestry might look similar to the untrained eye, but usually consists of rows of trees of the same species and age, planted close together so they shade the forest floor, which discourages the growth of other vegetation. It is effectively a monoculture like any other crop, not very different from a barley field. The trees may not even be native species to the country where they are grown. It can be a virtual desert from a biodiversity point of view.

However, as is the case with any monoculture of closely spaced individuals of the same species, a managed forest can be subject to infection, which can spread quickly from tree to tree, especially as the trees in forestry situations may be stressed or overcrowded. Some beetle species, especially of the families Curculionidae and Cerambycidae, can establish large populations in tree monocultures very quickly. These can be harmful to the survival of the trees themselves, or can affect the quality of the timber, making it less valuable when harvested.

BELOW | *Anoplophora chinensis* (Cerambycidae) An Asian species accidentally introduced into parts of Europe. The related Asian Longhorn (*A. glabripennis*) is a pest in North America.

LEFT | *Agrilus planipennis* (Buprestidae) Emerald Ash Borer, an introduced pest from China, emerges from an infested ash tree in the USA.

BELOW | *Ambosiodmus lecontei* (Curculionidae) From North America, this female has laid eggs and is cultivating white fungus for the larvae to eat.

Beetle infestation is more serious if the beetle is a vector of a tree pathogen, and such tree diseases can have serious consequences that can also affect trees grown away from monocultures. The most famous example is the Dutch elm disease fungus *Ophiostoma novo-ulmi*, which is spread by elm bark beetles of the genus *Scolytus* (Curculionidae). The female bark beetle carries fungal spores in special organs called mycangia, and inoculates them into the tree in which she lays her eggs, so that the larvae developing in their tunnels can feed on the fungal fruiting bodies. In the case of Dutch elm disease, the fungus being transmitted by the beetles became extremely pathogenic to the trees, and killed millions of elms in Europe, Asia, and North America.

Insect pests can be much more damaging outside their native range, where the local trees of the genus that they attack have not evolved alongside them, and where specific parasites and predators that usually keep the pest numbers under control are absent. Examples include forestry pests and pests of amenity planting such as street and garden trees, like *Agrilus planipennis*, the Emerald Ash Borer (Buprestidae), and *Anoplophora glabripennis*, the Asian Longhorn (Cerambycidae), both much more problematic in North America than they are in their native China.

Under adverse climatic conditions, even a native species can become a very serious forestry pest. The Mountain Pine Beetle *Dendroctonus ponderosae* (Curculionidae), for example, has been responsible for the death of millions of acres of pine forest in North America, peaking in 2009, after a series of dry summers placed trees under environmental stress. This, coupled with mild winters, failed to reduce the beetle populations to manageable levels.

DOMESTIC PESTS

With at least 400,000 species, utilizing most of the terrestrial habitats of the world, it is not surprising that a few beetles have adapted to take advantage of the opportunities offered by human-altered environments, especially considering the ever-increasing proportion of the earth's surface modified by humans. There are several beetle species that live in our homes and gardens, eating the food we store for ourselves or the plants we cultivate, or even the structure of our houses. These beetles are called "synanthropic" species (from the Greek "with people"), but when they come into conflict with us they are usually just called pests. Here, we use "domestic pests" to distinguish those of houses and gardens from those of agriculture and forestry, which are discussed elsewhere (see pages 54–59).

Every garden and almost every human dwelling is shared with at least a few beetle species, and archeology and paleontology show that this was always the case. Wherever people go, we carry crops and domestic animals, along with soil and animal food and bedding, and all these things carry beetles. Beetles are nature's recyclers, and species that have adapted over millions of years to a certain substrate often don't differentiate between a dead branch on a tree and one that has been carved into a piece of furniture, or animal remains left by predators and skins or meat prepared for human use.

As with synanthropic vertebrates such as rats, mice, and pigeons, it is often the same relatively few species of beetles that we have carried around the world over the centuries with migration and trade. In many cases, these species were with us for so long that their original geographical distribution is unknown. The beetles extracted from the mummies of the pharaohs of ancient Egypt are often the same species found in Viking archeological sites, in medieval burials, and as pests of dried meat products today.

The major limiting factor for beetles in a human dwelling is that it is very dry, since it never rains indoors. Many domestic beetle pests belong to the superfamily Bostrichoidea, which seem best able to cope with the dryness. These include the family Dermestidae, carpet beetles and larder beetles, which feed on dry animal protein such as woolen blankets and rugs, dry pet food, and dry meat and fish. Several genera such as *Anthrenus* and *Reesa* are called museum beetles, and will also attack taxidermy and insect collections, where they can be very destructive. Related beetles

of the family Ptinidae attack dry plant matter:
Woodworm *Anobium* and Death Watch Beetle
Xestobium rufovillosum can attack beams and
timbers, the latter earning their name from the
ticking sound they make at night to attract a mate.
The cellar beetles (genus *Blaps*) are large, black
beetles that can sometimes still be found in very
old buildings or stables, and in many European
cultures are connected with death; they are
also strongly nocturnal, and in both cases the
superstitions probably arose because rural people
before electricity, sleeping at dusk and rising at
dawn, will only have encountered night-active
insects in times of sickness.

POLLINATION

Angiosperms, flowering plants, are the dominant plant group in land ecosystems today, in both volume and species diversity. Most of the crops we eat are angiosperms, and as well as feeding us and shaping the habitats of the planet, plants produce the oxygen we breathe. Humans, like beetles and most other animals, rely on plants to exist.

As land plants are stationary, they cannot search for mates like animals do, so to reproduce, they discharge their male gametes, pollen, into the environment, aiming for some to reach the female flowers of another plant of the same species. Many plants rely on wind pollination, producing huge quantities of pollen so that a few grains of the millions will reach the right female flowers, but a more targeted strategy, in many of the higher plants, is to use insects for pollination.

Insect-pollinated plants have obvious flowers, combining both male and female organs, often developing at different rates in each flower to avoid self-fertilization. They provide a reward to attract pollinating insects—usually sugary nectar—and to access it, the insect must contact stamens that deposit pollen and a stigma that removes and uses any pollen from other flowers. Insects move between flowers of the same species, searching for the reward, and at the same time distributing the plants' genetic material. Some insect groups become specialized pollinators of particular flowers, guaranteeing a source of food, and at the same time ensuring the plants reproduce. Such insects often evolve structures that make them better pollinators, such as the fuzzy setae of bumble bees and flower chafer beetles that hold more pollen. This in turn increases populations of the plant on which they rely, so is an example of mutualism, a relationship between organisms where both benefit.

When we think of pollinators, we often think of bees and butterflies, but in the early Cretaceous when flowering plants first diversified, beetles were likely the first pollinators, just as they were, and still are, the primary pollinators of cycads and some other ancient plants. In modern ecosystems, beetles still play an important role in pollination, especially of flowering trees, and in tropical and arid environments. Beetle-pollinated flowers are often large or clustered, pale-colored, and may have a strong, sweet, sometimes even sickly scent. Examples include water lilies, arum lilies, magnolia, and hawthorn blossom. The fleshy petals of many of these, especially magnolia and water lilies, reflect an ancient association with beetles and protect the flowers from the beetles' chewing

LEFT | *Meligethes aeneus* (Nitidulidae) A mass of tiny pollen beetles pollinate a yellow zucchini (courgette) flower in an English garden.

ABOVE | *Anchylorhynchus* (Curculionidae) South American palm-pollinating weevils—similar African weevils are used to pollinate commercial oil palm.

BELOW | *Eupoecila australasiae* (Scarabaeidae) A fiddler beetle pollinating an Australian eucalyptus. Although these beetles eat part of the flowers, the benefits outweigh the damage.

mouthparts. Some beetle pollinators damage the plant, eating the pollen and stamens, or developing as larvae inside the ovules, so the pollination service has a cost.

Pollinators are worth billions of dollars annually to agriculture. Some important crops have failed to thrive in parts of the world because of a lack of native pollinators (hand pollination is far too labor-intensive). A famous example is the introduction of West African oil palm to tropical Asia, which has only been possible due to the introduction of pollinating weevils (genus *Elaeidobius*). This is a bittersweet success story, because oil palm has been so destructive to Asia's rainforests.

BIOLOGICAL CONTROL OF WEEDS

As people traveled around the world they brought with them exotic plants, sometimes to remind them of home, or as crops, ornamentals, food for livestock, to modify the landscape in some way, or even by accident as seeds in animal feed or bedding. Some of these plants then established and proliferated in the new environment and became invasive weeds. Unchecked by the usual herbivores and competitors, these plants can grow faster and denser than usual, crowding out native habitats and becoming severe pests, to the extent that invasive species have been described as the second biggest threat to biodiversity after habitat destruction.

Often, when a plant is established in a new environment, controlling it by the usual methods becomes impossible, as it establishes a seed bank in the soil—and it is then that people generally resort to biological control. The object of biological control is straightforward. The invasive plant has become so successful because it has escaped the usual checks and balances that regulate its population in its native range, so scientists look at the wild population where the target plant originated, to seek specific herbivores and seed predators, and import them as well, in the hope that they might regulate the pest without affecting anything else in the environment. This has to be carefully managed, since the wrong "biocontrol agent" might become an invasive species in its own right, or switch hosts and attack a vulnerable native relative of the target plant. Nowadays, a lot of laboratory testing takes place before biocontrol agents can be released, and only agents shown to be host-specific to the target weed are selected, to reduce collateral damage to other plants. Beetles, as well as Lepidoptera (butterflies and moths), are most frequently selected against invasive plants, because many of them are very host-specific, feeding on only one genus or species of plant. Among beetles, weevils (Curculionoidea) and leaf beetles (Chrysomelidae) have been used most often. Successful examples include Thistle Weevil *Rhinocyllus conicus* (Curculionidae), introduced from Europe to Canada, and Gorse Weevil *Exapion ulicis* (Brentidae) from Europe to New Zealand.

LEFT | Larvae of *Oxyops vitiosa* (Curculionidae) An Australian weevil biocontrol agent for paperbark tree, an invasive pest tree in Florida's wetlands.

Biological control of plants has generally been much more successful than biological control of animals, because most plant feeders are much more specific in their diet than most predators. A notorious historical failure of biological control of animals was the introduction of the South American Cane Toad into Australia to control the sugarcane beetle *Dermolepida albohirtum*. The toad was a totally inappropriate biological control agent, since it feeds indiscriminately on a wide range of insects, arthropods, and even small mammals, reptiles, and amphibians. Furthermore, it is toxic, and therefore dangerous to native predators such as monitor lizards and snakes that try to eat it. It remains a serious invasive in Australia to this day; it was also not very effective at controlling the sugarcane beetle.

ABOVE | *Agasicles hygrophila* (Chrysomelidae) The Alligator Weed Flea Beetle was introduced to Florida to control the invasive alligator weed that was blocking waterways.

RIGHT | *Nanophyes marmoratus* (Brentidae) A European weevil successfully introduced to North America to control the flowering plant European purple loosestrife.

SOURCE OF FOOD

Insects have probably always formed part of the human diet, and still do in most parts of the world today. It is only in Europe and North America that people have generally turned their backs on "entomophagy," as the eating of insects is called. In the Bible Lands at the time of the writing of the Old Testament, the eating of certain insects was explicitly encouraged (while mollusks and crustaceans were considered unclean).

Probably in more northern countries, where insects are few, small, and seasonal, and thus difficult to harvest in quantity, while marine invertebrates such as shellfish, crabs, and lobsters are large and abundant all year round (and stay fresh longer in the cooler climates), these preferences were reversed, and now insect eating seems strange and unpleasant to many Europeans and North Americans who would not think twice about a plate of oysters or shrimp. However, insects breed rapidly, can be fed cheaply, require

LEFT | *Rhynchophorus ferrugineus* (Curculionidae) Palm weevil larvae fried with garlic and chili on a bed of rice, served as a bar snack in southern Thailand.

ABOVE | *Tenebrio molitor* (Tenebrionidae) Usually ground into flour, but here entire mealworm larvae are served as an alternative to meat.

very little land, and produce very little carbon dioxide per pound of protein produced, so they represent a sustainable alternative to almost all other meat sources, and may have an important role to play in the future in feeding a growing human population.

As in most things, beetles feature largely in the list of edible insects, especially the larvae, which are generally bulkier than the adults, easier to collect in quantity, and lack the hard chitinous exoskeleton. The larvae of weevils (Curculionoidea) and scarabs (Scarabaeoidea) are among the most popular; they contain no noxious chemicals, can

BELOW | *Oryctes rhinoceros* (Scarabaeidae) Turning a pest into a meal. These large larvae destroy palm trees but are rich in protein. They look more appetizing after cooking.

be pests, and may occur in large numbers on agricultural land. In many tropical cultures, the pith of palm trees, such as sago, is used as a source of starch and carbohydrate, while the larvae of palm weevils (Curculionidae: *Rhynchophorus*), which develop in the same substrate and may be encountered while harvesting the sago, are used as a protein supplement. In other cultures, grubs of chafers (Scarabaeidae: Melolonthinae) that occur in the soil developing on the roots of crops may be collected in large numbers and either used to feed poultry or cooked and eaten directly, thus combining food foraging with pest control.

In Western countries, mealworm larvae (Tenebrionidae: *Tenebrio*) are grown in industrial quantities, not just as pet food but also to be ground up into "insect flour," which is used to make protein-rich breads or energy bars. Whole dried and spiced larvae may also be sold as an artisanal condiment, and recently we have seen a gradual normalization of insects as a food source in Western cuisine.

Adult insects are eaten less commonly, but in the night-markets of Southeast Asia, one can buy a large selection of flash-fried adult beetles straight out of a wok, which include water beetles (Dytiscidae and Hydrophilidae) and chafer beetles (Scarabaeidae: Melolonthinae), as a crunchy midnight snack. These are usually attracted to lights, gathered, and fried immediately.

ART

Since the dawn of representative art, people have drawn the objects and creatures with which they share their environment. While large prey animals dominate the earliest cave paintings and rock art, there are also some representations of insects, sometimes interpreted as beetles. Occasionally, beetles have featured in mythology, or have gained allegorical significance. A famous example, discussed on pages 72–73, is the ancient Egyptian veneration of scarab beetles, paintings and carvings of which adorn the pyramids and sarcophagi of the pharaohs.

There was a revival of such scarab imagery in the 1920s in Europe and the USA, when the discovery of Tutankhamun's tomb in 1922 led to a surge of interest in Egypt. The Art Nouveau and Art Deco movements incorporated stylized scarab motifs into art, jewelry, and architecture, and Egyptian-style scarab beetles, often with wings outstretched, are a common feature of buildings of that era. Today, visitors to London Zoo can see a much more realistic scarab sculpture, a modern bronze statue of a pair of *Kheper* (Scarabaeinae) with a dung ball, emphasizing the essential role dung beetles play in the African savanna.

Beetles feature prominently in still-life paintings of the Dutch Golden Age (1609–1713), particularly those of the Vanitas style, where symbolic objects remind viewers of their mortality and the fleeting nature of worldly goods and pleasures. Stag beetles *Lucanus cervus* were most frequently used, and receive a symbolism similar to that of the beetle's evening flight in literature, representing the end of day and, by extension, the end of life. An opulent table laid with food and drink was often juxtaposed with a single dead beetle, skull, or snuffed-out candle, as a *memento mori*. Stag beetles were probably popular because of their impressive size and menacing jaws, and it is consistent with the theme of the paintings that they were obviously dead.

Albrecht Dürer painted his famous stag beetle in 1505; at a time when nature was little valued, such a choice of subject was in the true spirit of the Renaissance. Although defiantly and vibrantly posed, the articulation is slightly unnatural, suggesting the model was dead. The use of actual dead beetles in art has a long history, especially in India and Indochina, where beetle elytra decorate hangings and muslin dresses. This inspired the

LEFT | Illuminated page from *Model Book of Calligraphy* (1561–96) by Bocskay and Hoefnagel, showing an accurate European rhinoceros beetle *Oryctes nasicornis*.

famous "Beetle Wing Dress" worn by Victorian actress Ellen Terry in her portrayal of Lady Macbeth (1888–89), now preserved at her home at Smallhythe Place, in Kent, UK. The same jewel beetles *Sternocera aequisignata* (Buprestidae) were used in large installations by Belgian artist Jan Fabre (born 1958), including "Heaven of Delight," where a whole room of a Brussels Palace, including the chandeliers, is encrusted with thousands of elytra. Another metallic green beetle often used in jewelry is the tortoise beetle *Polychalca punctatissima* (Chrysomelidae) from Brazil. These were popular in the 1920s and were often sold, in the spirit of the times but quite wrongly, as "scarabs." Sadly, they seem much rarer now. Many of the Atlantic coastal forests where they used to live have been destroyed.

LITERATURE

In his "Ode on Melancholy" (1820), the romantic poet John Keats urges the reader: "Make not your rosary of yew-berries, Nor let the beetle, nor the death-moth be, Your mournful Psyche." The admonition is not to indulge too freely in baleful associations of death. "Psyche," representing the human soul, is often depicted as a butterfly. The "death-moth," on the other hand, refers to a quite different and more nocturnal lepidopteran, the Death's Head Hawk Moth, a large insect with a pattern on its thorax that resembles a human skull, and which has long been a literary and artistic symbol of mortality. "Yew-berries" are deadly poisonous, and yew trees with their dark foliage are traditionally planted in churchyards. But why beetles? The noisy, bumbling, crepuscular flight of beetles such as Geotrupidae, Cerambycidae: Prioninae, and Lucanidae has long symbolized the end of the day and the beginning of the night, and they have become literary guardians of the darkness, like moths, inhabitants of the veil that separates the brightness of day from the darkness of night (and by extension, life from death). The same use of the beetle's flight is seen in the earlier "Elegy Written in a Country Churchyard" (1751) in which the poet Thomas Gray describes the evening: "Now fades the glimm'ring landscape on the sight, And all the air a solemn stillness holds, Save where the beetle wheels his droning flight." The beetle is the only thing breaking the graveyard's silence. Shakespeare's Macbeth (1606) uses the same image and other images of darkness to confess the murder he is planning: "Ere the bat hath flown His cloistered flight, ere to black Hecate's summons The shard-borne beetle with his drowsy hums Hath rung night's yawning peal, there shall be done A deed of dreadful note" (what he means is "before dark"). "Shards" refers to the

ABOVE | Bewitching splendor: Ellen Terry's famous Lady Macbeth dress was made from tropical Buprestidae elytra, slightly out of place in eleventh-century Scotland.

beetle's elytra, and its darkness is augmented by association with "black Hecate," the queen of witches.

These three examples span more than 200 years, but an entomologist is immediately struck by how commonplace large, evening-flying beetles must have been to the writers, and their audiences, to acquire such symbolism. Big beetles such as *Geotrupes* are no longer very frequently seen or heard, but in an era before electric-light pollution, agricultural intensification, and pesticides, they must have been, in season, familiar and everyday harbingers of twilight.

Other literary uses of beetles are more diverse. Wordsworth (in 1802) discusses their detail under magnification: "The beetle panoplied in gems and gold, A mailed angel on a battle-day," and Gregor Samsa, the unfortunate protagonist of Franz Kafka's *Metamorphosis* (1915), wakes up to find himself transformed into a gigantic insect, which is often interpreted or illustrated as a beetle. More recently, M.G. Leonard's popular Beetle Boy trilogy of children's books follows the adventures of three children, Darkus, Virginia, and Bertolt, and their beetle companions, Baxter, Marvin, and Newton, as they try to rescue the Natural History Museum director and save the world from eco-terrorism.

ABOVE RIGHT | A stag beetle from the 1491 medieval treatise *Hortus Sanitatis* (Garden of Health), which attributes dubious medical properties to its mandibles.

RIGHT | *Battle of the Beetles*, the final book in The Beetle Boy Trilogy. In modern children's literature, the insects are often characters rather than just background.

71

MYTHOLOGY

Although beetles and other insects feature widely in the folklore of many parts of the world, the dung beetle is one of the only insects to have been elevated to the status of a god! The scarab-headed god Khepri was one of the manifestations of the chief god Ra in the ancient Egyptian pantheon, and was responsible for the morning sun. The ancient Egyptians, a largely pastoral people, would have watched the behavior of scarab dung beetles while out tending their herds, and the dung beetle rolling its ball of animal dung across the pasture became a metaphor for the sun god rolling the flaming ball of the sun across the sky. Khepri is usually shown with a whole scarab beetle for a head; he was also the god of renewal and rebirth, because he was believed to have been formed from "nothing," just as the adult dung beetles appear to hatch perfectly formed from their dung balls, without any obvious immature stages. The connection between an adult beetle seemingly emerging from waste and taking flight with rebirth after death is intuitive, and carved scarab amulets are one of the most common surviving artifacts from most periods of ancient Egypt, dating back to the Bronze Age.

The dung beetle most commonly represented in ancient Egyptian art is the Sacred Scarab *Scarabaeus sacer* (Scarabaeidae), named because of this association, but several other genera and species are depicted, including members of the genus *Catharsius*, the single-horned *Copris*, and the African genus *Kheper*, whose name is a variant of the name of the god. Clay models of the genus

BELOW | A vast Egyptian scarab beetle statue, six thousand years old. Carved from a single block of diorite, it is thought to represent the god Khepri.

Copris are also common archeological finds in Minoan Crete, although it is not known to what extent these are connected.

The ancient Egyptians are not the only people to have incorporated beetles into their creation mythology. The Cochiti people of the southwestern US have a story about the desert beetles of the genus *Eleodes* (Tenebrionidae), which raise their abdomen when they are alarmed to release a defensive secretion. The Cochiti tell that the beetles are hiding their face because long ago they had the task of arranging the stars in the sky, and they dropped them, leaving the randomly scattered pattern we see today. They are so ashamed of this mishap that when they hear someone approach, they hide their head in the dirt.

73

POPULAR CULTURE

Collecting and studying beetles, or other objects from nature such as plants, fossils, and butterflies, was a popular pastime in various parts of the world at different times, when people had sufficient leisure, education, and access to natural habitats. Sometimes this is a background hobby with only a few adherents in any generation, but sometimes it becomes fully integrated into the popular culture of a time and place, and when this happens, people learn from one another, collect together, publish guidebooks, and interest in nature becomes more widespread, until a more "nature literate" society emerges. An early example of such a flowering is Victorian and Edwardian Britain, where butterfly nets or geological hammers were a standard part of many affluent people's childhoods. The acceptance and encouragement of natural history-related hobbies at this time ultimately led to the world being enriched by the theories of Charles Darwin, Alfred Russel Wallace, and Henry Walter Bates; the discoveries of Mary Anning and Gideon and Mary Mantell; and the foundation of great museum collections, which are largely distillations of private collections of previous generations of amateurs. Private collections have been described as streams that flow into and feed the rivers of institutional collections, and ultimately into the sea of human knowledge.

Other notable occasions when entomological interest has blossomed and become integrated into popular culture have been in Europe, notably the former Czechoslovakia, Poland, East Germany,

LEFT | *Allomyrina dichotoma* (Scarabaeidae)
A girl admires a Japanese rhinoceros beetle, called Kabutomushi (helmet beetle) for the impressive forked horn that resembles a samurai helmet. The beetle remains common in Japan.

and Italy, in the mid-late twentieth century, and in eastern Asia, initially Japan and later Korea, China, and Taiwan in the later twentieth century to today.

The integration of entomology into popular culture is impeded by urbanization and the loss of convenient access to natural habitats, effectively forcing physical separation of people from nature. Japanese game designer Satoshi Tajiri, creator of the Pokémon games franchise, was an enthusiastic insect collector as a boy, and observed that as urban areas spread and land was paved over, habitats where he hunted insects were lost. Tajiri has said that he wanted his games to allow children to "have the feeling of catching and collecting creatures as he had." Pokémon has become an international and cross-generational sensation, and Japan still produces many amateur and professional entomologists as well.

Another threat to entomology in popular culture is when the disconnect between people and nature leads to misguided "conservation" strategies, not protecting habitats and environments from destruction, but instead protecting individual insects and plants from collection and study, which ironically makes nature less relevant to the population, and therefore more at risk in the long term. Veteran broadcaster and conservationist Sir David Attenborough has spoken out against this tendency in the British newspapers, stating, "Children are being denied the chance to learn one of the key 'foundation stones' of science [that is, taxonomy] because of laws that prevent them from collecting wild flowers, insects, and fossils." His powerful words will hopefully help children to develop an interest in and desire to protect the natural world around us.

BIOMIMICRY

The famous British naturalist, author, conservationist, and founder of Jersey Zoo, Gerald Durrell (1925–95), was fond of explaining, especially on long sea journeys returning from collecting trips, that everything invented by humans had already been invented beforehand by animals. He recounts in his memoirs that his audiences rarely believed him, so he would back up his assertion with examples of sonar used by bats and whales, aqualungs and diving bells by water beetles and water spiders, electricity by electric eels, and so on. There is little doubt that many inventions were inspired by careful observation of nature, nor that humans and other creatures need to overcome similar problems in order to survive, whether by adaptation or by technology. The new science of biomimetics is simply a formalization of a process that goes back to the dawn of humanity, a deliberate attempt to search the natural world for solutions to problems, chemical substances, and physical designs that can be replicated or

ABOVE | *Chrysina gloriosa* (Scarabaeidae) The brightly colored exoskeleton of this scarab beetle consists of thousands of prisms, and could be copied in order to make reflective surfaces.

BELOW | *Onymacris unguicularis* (Tenebrionidae) The Head-Stander Beetle is a fog basker, meaning it is able to condense minute atmospheric water droplets in the dry Namib desert.

harnessed for a similar purpose. This can result in the refinement of existing technologies or the development of new ones. Since beetles are so diverse, and have adapted to so many different habitats and ways of life, they are an obvious place to search.

One of the greatest breakthroughs of this kind (although not linked to beetles) was the discovery of penicillin from a toxin produced by mold fungi to kill competing bacteria, which led to a wide range of antibiotics. Many carrion-feeding beetles have a similar need to prevent microbial activity, for example burying beetles of the genus *Nicrophorus* treat the carcasses on which they rear their larvae with antibacterial substances, to slow down decay. Knowing the biology of beetles, and which ones actively compete for resources with bacteria or fungi, may indicate a fruitful place to search for new generations of antibacterial or antifungal substances. Recent studies have synthesized a chemical based on harmonine, a defensive secretion of a ladybug. This chemical was shown in the laboratory to be effective against some agents of human diseases, such as protozoans (including *Leishmania* and *Plasmodium*, which cause leishmaniasis and malaria) and some bacteria (including mycobacteria, which cause leprosy and tuberculosis).

As well as antimicrobial chemicals, some large Staphylinidae and Silphidae also secrete substances that are very unpleasant to vertebrate scavengers such as dogs, to prevent them from eating the carrion in which the beetles are breeding. Such substances might potentially be copied to make deterrents for mammals, for example to keep foxes away from a chicken house.

Insects that have conquered the most inhospitable environments on earth are a rich seam of models for biomimetics. Beetles in the Namib desert survive the dry conditions by fog basking: harvesting water vapor from the atmosphere during brief fogs and condensing it on their exoskeleton. Through biomimetics, synthetic surfaces that imitate this complex structure can extract water vapor from the atmosphere, a technology that can be used for self-filling bottles in arid regions, or removing excess humidity from places where it is not wanted, such as inside electronics.

MEDICINE

Compared to many groups of insects—such as Hymenoptera (ants, bees, and wasps) that can bite and sting, and Diptera (true flies), which include vectors of important diseases, such as malaria and yellow fever carried by mosquitoes—beetles are relatively innocuous from a medical and human health point of view. Blistering of human skin can be caused by exposure to the cantharidin secreted by several beetles of the families Meloidae and Oedemeridae, the former often called "blister beetles" and the latter, in parts of Oceania, "bubble bugs" because of the circular, liquid-filled blisters they can cause, for example if accidentally crushed against the skin. The merits of the Spanish Fly *Lytta vesicatoria* (Meloidae), a green beetle from southern Europe, as an aphrodisiac has been discussed for centuries, including by the notorious Marquis de Sade (1740–1814)—whose name is the origin of the word "sadist"—but it is, in fact, a deadly and painful poison.

Blisters can also be caused by exposure to some Staphylinidae of the genus *Paederus*, which includes several hundred species worldwide, such as the Afrotropical Nairobi Eye Fly *Paederus eximius*. These brightly colored flying insects are attracted to light and may enter houses as a result. They are also called "dragon bugs" because of their bright colors and the burning sensation that they cause on contact with the skin. The active ingredient in this case is not cantharidin but a similar chemical, pederin, which is extremely rare in nature (apart from these beetles it is only known from some marine sponges) and is under investigation as a potential anticancer drug.

Other drugs derived from insects are also under investigation. The light-producing enzyme luciferase and associated genes, used in various Elateroidea, such as fireflies, for mating signals or in defense, have been applied medically for monitoring the progress of infections or to visualize, for example, liver cells in human embryonic development. Some beetles such as the Yellow Mealworm *Tenebrio molitor* (Tenebrionidae), because they can be bred in enormous numbers—up to several tons per day

in suitable facilities—can be modified to express virus antigens and used as an affordable and scalable bioreactor for mass antigen generation for use in vaccine production against emerging viruses. The use of the ladybug *Harmonia axyridis* (Coccinellidae) to produce the antimicrobial agent harmonine has been discussed under biomimicry (see page 77). Beetles are likely to yield dozens or hundreds of useful substances if the technology and the will to study them are available.

Beetles feature in traditional medicine in some parts of the world. In Southeast Asia, whirligig beetles (Gyrinidae) are collected from ponds and used as a folk remedy for fever. In traditional Chinese medicine, fireflies of the genus *Luciola* were once thought to clarify eyesight, cure night blindness, and treat wounds and burns caused by fire. Finally, some beetles, especially tenebrionids that live around human habitation (genera *Blaps* and *Ulomoides*, for example), may be intermediate hosts for tapeworms or roundworms that can affect humans.

ABOVE | *Lytta vesicatoria* (Meloidae) The Spanish Fly is a source of cantharidin, which may have medical applications but is toxic if consumed.

OPPOSITE | *Paederus fuscipes* (Staphylinidae) This widespread and brightly colored rove beetle secretes the vesicant pederin, which has been tested for anticancer properties.

BELOW | *Photinus pyralis* (Lampyridae) The Big Dipper Firefly of North America—its bioluminescence chemicals are helpful for marking growing cells.

BEETLE CLASSIFICATION

Our system for classification of beetles dates back to the Swedish physician and botanist Carolus Linnaeus (1707–78), who published the basis for zoological nomenclature in the tenth edition of his *Systema Naturae* (System of Nature). Linnaeus recognized what he considered to be distinct species, and gave each a Latin-based binominal name consisting of a genus, which it shared with other similar creatures, and a species name that was unique within each genus. The result was that every species received an exclusive name, but was also in a genus with similar creatures. We would, of course, say "related" today, and it is difficult to remember that Linnaeus was working on imposing order on the astonishing diversity of life, without any evolutionary framework, and that his classification predates the publication of Darwin's *Origin of Species* (1859) by more than a hundred years. He was recognizing similarity without the philosophical tools to comprehend the reasons for it.

Linnaeus's visionary nature is reflected in the fact that we still use his system today, and the "Scientific Names"—also called "Latin Names" or "Latin Binominals"—are regulated by the International Commission on Zoological Nomenclature, a body of distinguished scientists from around the world that is currently based in Singapore. Linnaeus's importance was recognized by his contemporaries. Indeed, the philosopher Jean-Jacques Rousseau sent him the message: "Tell him I know no greater man on earth," while Goethe wrote "With the exception of Shakespeare and Spinoza, I know no one … who has influenced me more strongly."

Linnaeus's genius was in devising a simple system that could be extended and was effectively

future-proof. He used Latin simply because it was the language of scholarship at the time, but this was ultimately fortuitous, because it meant that scientists could communicate about organisms using a universal language.

Subsequent generations, starting with Linnaeus's students and followers, added more taxonomic ranks, which became like a system of nesting boxes, each one (as one goes up the classification) more inclusive than the one before. Ranks above genus are distinguished by their endings, and sometimes called the "Higher Classification." There are conventions for how Scientific Names are written, for example the genus and species are written in italics, the genus name always has a capital letter, the species name never does (even when based on someone's name).

The European Stag Beetle would be classified, in descending order, as:

1. **Kingdom Animalia (animals)**
2. **Phylum Arthropoda (arthropods, animals with jointed limbs)**
3. **Class Insecta (insects, arthropods with three divisions of the body and six legs)**
4. **Order Coleoptera (beetles)**
5. **Family Lucanidae (stag beetles)**
6. **Genus *Lucanus* (typical stag beetles)**
7. **Species *Lucanus cervus* (European Stag Beetle)**

This classification system can easily be modified or extended, which to some extent explains its survival. For example, extra ranks can be inserted above or below existing ones simply by adding "super-" or "sub-" to the rank name. So, the stag beetle family Lucanidae can also be placed in the superfamily Scarabaeoidea, for scarablike beetle families.

In this book, we use the classification included in a chapter on global beetle diversity published by one of the co-authors in 2017.

PHYLOGENETIC RELATIONSHIPS

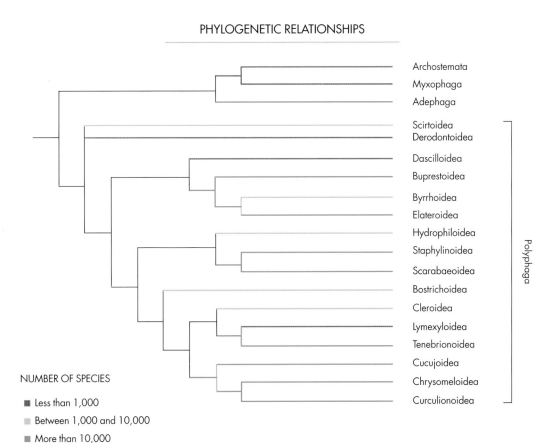

Archostemata
Myxophaga
Adephaga
Scirtoidea
Derodontoidea
Dascilloidea
Buprestoidea
Byrrhoidea
Elateroidea
Hydrophiloidea
Staphylinoidea
Scarabaeoidea
Bostrichoidea
Cleroidea
Lymexyloidea
Tenebrionoidea
Cucujoidea
Chrysomeloidea
Curculionoidea

Polyphaga

NUMBER OF SPECIES

■ Less than 1,000
▨ Between 1,000 and 10,000
■ More than 10,000

ARCHOSTEMATA

Archostemata is the smallest of the four suborders of Coleoptera, with fewer than 50 known living species. It is an ancient lineage with a rich fossil record dating back at least to the Triassic, before the age of the dinosaurs, and living species preserve the simplified appearance of some of the earliest beetles. Although there are very few species, the modern fauna has five families, Jurodidae, Micromalthidae, Crowsoniellidae, Ommatidae, and Cupedidae, but the first three are each known from only a single living species. Two of these are among the most mysterious and little-understood of all insects: Crowsoniellidae was collected only once when three males were found deep in the soil of Italy in 1973; Jurodidae are known in the modern fauna from a single specimen collected in the forests of Sikhote Alin, Russian Far East. Neither was ever seen again, despite several expeditions. Other families are better known, but are still generally rare. Ommatidae, with three genera, *Omma*, *Beutelius*, and *Tetraphalerus*, is restricted to Australia and southern South America. Species of Cupedidae, the least rare of the families, can occasionally be found in all continents except Europe and Antarctica.

The few species of Archostemata that remain in the world today closely resemble fossils found in stone and amber from hundreds of millions of years ago, showing that they have changed little since the Paleozoic era. Of the five living families of these ancient relicts, three are now relatively well known: the small and rather bizarre Micromalthidae and the larger Cupedidae and Ommatidae.

Micromalthidae, tiny, wood-feeding beetles from North America, have one of the strangest life cycles in the animal kingdom. They have 'pedogenetic' larvae, that is larvae that are themselves able to produce more larvae as offspring. The secondary larvae then eat their way out of the mother larva, before molting and developing into other mother larvae and fertilizing themselves—or, rarely, pupating and becoming an adult. The adults are an evolutionary dead end, as they apparently have no functioning reproductive organs.

Cupedidae and Ommatidae, families quite similar to each other, lack the peculiarities of

SUBORDER
Archostemata

KNOWN SPECIES
40

DISTRIBUTION
Scattered distribution with a few species known from all continents except Antarctica

HABITAT
Mainly in forest habitats, in and around living and fallen trees

SIZE
2–30 mm

DIET
Larvae of most species, where the larval diet is known, feed on fungi in dead wood or roots. Adults may feed on pollen or tree sap, but in some cases hardly feed at all

NOTES
The only known living species of Micromalthidae is probably common in North America, and has spread to

ABOVE | *Micromalthus debilis* (Micromalthidae) A rare picture of an adult *Micromalthus*, or Telephone-pole Beetle, among larvae. Strangely, it is the larvae that will reproduce; the adult is not capable.

BELOW | *Omma stanleyi* (Ommatidae) at rest on dead wood in Australia; this is probably the most primitive-looking living member of this ancient lineage of beetles.

Micromalthidae and are much larger, but they resemble no other group of beetles. Four species of Cupedidae, reticulated beetles, can be quite easily found in North America. In western USA, males of *Priacma serrata* are known to be attracted in numbers to some brands of laundry bleach, which apparently smells like the pheromone produced by the female. In Asia, Africa, Australia, and South America, only a few species of Cupedidae are known, and in Australia and South America, some Ommatidae, but none are ever common, and finding them is usually a matter of great persistence or good luck.

OPPOSITE | *Rhipsideigma raffrayi* (Cupedidae) imitates a bird dropping on a Madagascan rainforest leaf. This mimicry is interesting, since cupedids are a much older lineage than birds!

other continents with trade. Adults are hardly ever seen, as the larvae are able to reproduce without them. As beetle taxonomy is mainly based on adults, the species was considered rare. This suggests that other species of Micromalthidae may well exist undiscovered in less populated parts of the world, since tiny larvae in wood do not always attract attention. For this reason, DNA examination of timber may reveal new, unknown Micromalthidae

OPPOSITE | *Satonius stysi* (Torridincolidae) Larva of this remarkable species, discovered in 2007 on the Jade Dragon waterfall in Yunnan, China, and named in 2008 after the distinguished Czech entomologist Pavel Štys.

MYXOPHAGA

The suborder Myxophaga is the second smallest of the four Suborders of Coleoptera (after Archostemata) in number of species. It consists of two superfamilies, Lepiceroidea and Sphaeriusoidea: the first with just one family, Lepiceridae, and three species; the second with three families, Sphaeriusidae, Torridincolidae, and Hydroscaphidae, and a total of 102 described species. All species are small, none exceeding 3 mm, and most live in aquatic, semiaquatic, or hygropetric habitats, such as seepages, wet rocks under waterfalls, hot springs, and salt marshes. Myxophaga are rarely common, and although species are found in most continents, their small size and the rare habitats they frequent have left them poorly understood and rare in collections.

For such small beetles, Myxophaga have a rather long and detailed fossil record, suggesting an ancient group that was more abundant, diverse, and ecologically important in the geological past. Myxophagan fossils are known back to the Triassic period, more than 220 million years ago. One of the earliest fossils was initially mistaken for Staphylinidae, a much more recent family within the hyperdiverse suborder Polyphaga, but that placement has now been disputed.

SKIFF AND TORRENT BEETLES

All members of the suborder Myxophaga alive today are considered uncommon. Their small size, usually less than 2 mm long, also makes them difficult to find. One of the best techniques is to locate suitable habitats such as hot springs and seepages, and then search in detail for the beetles. This approach, especially when applied in areas where no Myxophaga have been reported before, has led to the discovery of several new species and genera in recent decades, and there are probably many more waiting to be discovered.

There are four families. Sphaeriusidae, minute bog beetles, includes 23 similar species of the genus *Sphaerius*, found on most continents. One species found in Britain, less than 1 mm long, is called *Sphaerius acaroides*, meaning "mitelike," which is a good name as these beetles can easily be overlooked as small, black oribatid mites that often occur in the same habitats. The skiff beetles (Hydroscaphidae; 22 species) are named for their boatlike shape, with pointed abdomen and short elytra, superficially resembling some Staphylinidae (rove beetles), but they are smaller than staphylinids, and unlike them, adults and larvae live on seepages or in mineral-rich springs grazing on algae. Torridincolidae (60 species) are

SUBORDER
Myxophaga

KNOWN SPECIES
102

DISTRIBUTION
Worldwide except Antarctica, common nowhere

HABITAT
Marginal habitats, including hot springs, waterfalls, snow melts, and the splash zone of mountain streams

SIZE
0.8–2.6 mm

DIET
Algae and other encrusting substances

NOTES
Recently, improvements to synchrotron technology have allowed scientists to access fossil insects preserved inside opaque substances, such as cloudy amber, and to scan and reconstruct their appearance. Recently, an international team applied this

technique to a coprolite, that is a 230
million-year-old lump of fossil excrement,
presumed to have been produced by a
Triassic dinosaur relative. It was shown to
be full of tiny beetles belonging to a new
genus and species of Myxophaga, now
named *Triamyxa coprolithica*. This not only
opens a window onto ancient ecosystems,
but also shows that amber is not the only
substance to allow detailed preservation
across huge expanses of geological time

called torrent beetles for their association with
fast-flowing streams and waterfalls, where they live
in the splash zone, not in the water itself. The three
species of *Lepicerus* (Lepiceridae) are found in leaf
litter in the neotropics, and the most recently named
was discovered in agricultural fields in Ecuador.

OPPOSITE | *Tricondyla* (Carabidae: Cicindelinae)
A wingless predatory tiger beetle found running on
the forest floor in Indonesia. These beetles mimic
large ants, a form of Batesian mimicry.

ADEPHAGA

The suborder Adephaga is the second largest of the four living suborders of beetles, with around 46,000 known species, although 40,000 of these are in one large family, Carabidae. Adephaga has a long fossil history, dating at least to the Early Triassic period, more than 240 million years ago, and by the Late Triassic, easily recognizable fossils of many modern families are found. Adephagans live in a wide range of habitats worldwide, from subterranean fissures to the rainforest canopy.

All Adephaga have glands on the abdomen that produce chemicals. In most cases these are used for defense, but they have other functions, including antifungal and antimicrobial, as well as modifying the aquatic habitat (such as breaking the water surface tension, or as a propellant in some Gyrinidae). These chemicals account for the characteristic smell of Adephaga, and staining of the fingers that may occur after handling carabids. Chemical defense is taken to an extreme in the carabid subfamily Brachininae (bombardier beetles) which can release a scalding hot cocktail of chemicals into the face of a predator, with an audible pop.

GROUND BEETLES, DIVING BEETLES, AND RELATIVES

The Adephaga is dominated by the very large family Carabidae, which has around 40,000 species worldwide. The other families are as follows, with numbers of known species: Gyrinidae (1,000), Trachypachidae (6), Rhysodidae (350), Haliplidae (220), Meruidae (1), Noteridae (250), Amphizoidae (5), Aspidytidae (2), Hygrobiidae (5), and Dytiscidae (4,000). Of these, Gyrinidae, Haliplidae, and Dytiscidae have been given their own sections below. Eight of the eleven families are mainly aquatic.

The family Meruidae was discovered in 2005, from waterfalls in Venezuela, and its name is based on an indigenous word for a waterfall. The tiny beetles cling to wet rocks, feeding on algae. Both members of the also recently described Aspidytidae show the same behavior. One species of aspidytid occurs in China, the other in Africa, a strangely

SUBORDER
Adephaga

KNOWN SPECIES
46,000

DISTRIBUTION
Worldwide except Antarctica. Some species live in the Arctic Circle

HABITAT
All terrestrial habitats

SIZE
1.5–100 mm

DIET
Predators, feeding on living invertebrates and sometimes even vertebrates. A few have become seed feeders or are known to graze on algae and detritus

NOTES
The name of the suborder has its roots in ancient Greek and means "gluttonous," for the extreme predatory habits of some members. Carabidae tearing apart a worm, snail, or even a small reptile, or Dytiscidae eating fish, newts, or water snails, is a

disjunct distribution. Another family that has a disjunct distribution is Hygrobiidae, known as screech beetles because of the loud squeaking sound they make as a defense when taken out of water, by pushing air through sound-producing organs. The five species are found in Europe, China, and Australia. Similarly, the trout stream beetles Amphizoidae are found in North America and China. Such distributions are probably evidence of an ancient group that has become extinct over most of its range but hangs on in a few seemingly random places.

Rhysodidae are relatives of Carabidae, and they are predators and fungus feeders in dead wood, while Trachypachidae are a small and obscure family that is found in forest litter in Eurasia, North America, and Chile.

OPPOSITE | *Clinidium* (Rhysodidae) A typical rhysodid, this beetle from Washington State is a predator in dead wood.

RIGHT | *Noterus* (Noteridae) Burrowing water beetles, this one from Europe, are common in leaf litter at the bottom of ponds.

memorable sight. Some, like the Caterpillar Hunters (Carabidae: *Calosoma*) have been widely introduced to control pest caterpillars. A few Adephaga have abandoned this voracious predation, and now feed on plant seeds, also rich in protein, or in a few cases even on algae

WHIRLIGIG BEETLES

The Gyrinidae is a small family of fewer than 1,000 species of small-sized beetles, found throughout the world, including Europe and North America. Despite their small size, they are well known and conspicuous, as adults can be seen in large numbers on the surface of still or slow-flowing water, moving in a very rapid, erratic, seemingly random motion, which has earned them the name "whirligigs," after old-fashioned spinning-top toys. They are predators, hunting on the surface tension of the water, and are constantly "reading" the ripples, which alert them to the presence of drowning insect prey.

Whirligigs are remarkable for having the eyes of the adults divided, one half observing the air above, the other watching for danger in the water below. They give the appearance of having four eyes,

BELOW | *Gyrinus* A European whirligig beetle on the surface tension, showing the upper part of the divided eyes and flaplike back pairs of legs.

FAMILY
Gyrinidae

KNOWN SPECIES
1,000

DISTRIBUTION
Worldwide except Antarctica

HABITAT
Still, fresh water such as ponds and lakes, as well as slow-moving rivers and streams. Some species live in brackish water such as rock pools

SIZE
3–15 mm

DIET
Larvae are active hunters below the water, while adults generally feed on drowned and drowning insects that have become trapped in the surface tension

but in fact the upper and lower halves are united within the beetle's head. Their divided vision, combined with their incessant frenzied movement, seems to protect them from being eaten by birds and fish, and as soon as one of them senses a threat, such as a shadow passing overhead, they all speed up their gyrations, or simultaneously dive to safety. This makes them very difficult to catch. They have small, paddlelike hind and middle legs, adapted for rapid swimming and changing direction. The predatory larvae live in the water, leaving only to pupate on emerging water-plant stems in the soil of the banks.

ABOVE | *Gyrinus* A larva with prey. The structures along the sides are gills for extracting oxygen from the water.

RIGHT | *Dineutus* An adult North American *Dineutus* on land, preparing for flight. Adults fly well between potential feeding and breeding ponds.

NOTES
Whirligig beetles are an ancient group of Adephaga, with reliable fossils going back to at least the early Jurassic period, almost 200 million years ago. It seems they were more species-rich and diverse in the past than now. Adults and larvae have been found preserved in stone fossils and amber from countries including Russia, Mongolia, Burma, Switzerland, and Germany, and most of these closely resemble the genera and species living today

TIGER BEETLES

The tiger beetles, a large and widespread subfamily of the ground beetles (family Carabidae) are, as their name suggests, formidable predators. The adults rely on their excellent vision and great speed—they are the fastest of all beetles, with the Australian *Cicindela hudsoni* having been clocked at an incredible 2.5 meters per second (equivalent to 5.6 miles per hour, or 125 body-lengths per second!). This speed enables them to hunt and capture fast-moving prey such as flies.

The larvae prefer to pursue a different hunting strategy, as sedentary ambush predators. They live in burrows, which they close with their flattened head, suddenly grabbing an unwary passing ant or

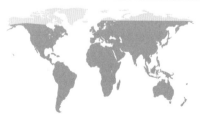

SUBFAMILY
Cicindelinae

KNOWN SPECIES
2,600

DISTRIBUTION
Worldwide except polar regions and Antarctica, most abundant in the tropics

HABITAT
Plains, savanna, lake and river margins, tropical forests

SIZE
6–70 mm

DIET
Adults and larvae eat other insects, especially ants and flies. Large species will tackle large spiders and small amphibians and reptiles

NOTES
Most tiger beetles are diurnal, visual predators with large eyes and bright metallic colors. Species that have become crepuscular or nocturnal can be recognized by their smaller eyes and duller colors

ABOVE | *Therates dimidiatus wallacei* Named after the explorer and evolutionary biologist Alfred Russel Wallace (1823–1913), this subspecies is photographed on a leaf in a forest park in Singapore.

OPPOSITE | *Manticora latipennis* One of the world's largest tiger beetles, seen here tackling a grasshopper in Gorongosa, Mozambique, Africa.

RIGHT | *Cicindela chinensis japonica* (Japanese Tiger Beetle) This subspecies of the widespread East Asian *Cicindela chinensis* is found in southern Japan, as far north as Tokyo.

beetle that comes within reach and dragging it into their tunnel.

Most tiger beetles are diurnal, and are characteristic insects of river margins and lake shores, sand dunes, and savanna, preferring warm places with well-drained soil for the larval burrows. Some species live in tropical rainforests, even in the tree canopy, where they hunt on leaves, flying rapidly from branch to branch in patches of sunlight.

The largest genera, such as the widespread *Megacephala* and the African *Manticora*, are flightless terrestrial hunters, and some have become nocturnal or crepuscular (active at dawn and dusk); these can be recognized by their more somber coloration compared to the bright metallic colors of their day-active relatives.

TYPICAL GROUND BEETLES

The huge and diverse subfamily Harpalinae accounts for almost half of the family Carabidae. Most people's view of a "typical beetle" is probably one of these. The subfamily shows some of the widest ranges of habits and habitats of any beetle group, but with a strong tendency toward predation. Some species are specific predators, attacking, for example, mainly snails, worms, or particular caterpillars, while others are generalists and will eat anything they can overpower. In northern Europe and North America, members of the genera *Pterostichus* and

Abax, large, black typical ground beetles may exist in enormous numbers on agricultural land, and some farmers will set aside small areas of uncultivated land along their crops as "beetle banks" to encourage these voracious and generalist predators to control slugs, snails, and larvae of pest insects in crops.

While most Harpalinae are beneficial, a few species are less welcome in crop fields—for example, the Strawberry Beetle (*Harpalus rufipes*) can be a pest of soft fruit by eating the achenes (external seeds) from growing strawberries, spoiling their

LEFT | *Craspedophorus* (tribe Panagaeini) A large nocturnal predator from Thailand. Many Carabidae are black with four distinctive red or orange spots.

OPPOSITE ABOVE | *Mormolyce phyllodes* (tribe Lebiini) A Guitar or Violin Beetle from Southeast Asia. The largest Harpalinae, at 4 in (10 cm), it lives among giant fungi.

SUBFAMILY
Harpalinae

KNOWN SPECIES
19,000

DISTRIBUTION
Worldwide except Antarctica, especially in the tropics, but also reaching into the Arctic Circle

HABITAT
Almost every habitat, from subterranean caves to the forest canopy

SIZE
2–100 mm

DIET
Almost all predators, though there are a few subgroups that are mainly seed feeders

NOTES
Some Harpalinae, for example in the tribe Lebiini, are parasitic as larvae, an unusual habit among beetles. In southern Africa, especially in the Kalahari Desert, the harpaline genus *Lebistina* is an ectoparasite

market value. Another harpaline that has become a pest of crops, such as wheat and barley, is the large, black *Zabrus tenebrioides*, which has adapted to feed on seeds. The adults climb up stems at night and eat the developing grain, often stripping the crop in steppe areas of Central and Eastern Europe. The larvae also feed on crops, often eating the sprouts. These pest species, however, are the exception.

ABOVE | *Anthia thoracica* (tribe Anthiini) From tropical Africa, this beetle can overcome large prey, including small reptiles, with its sharp, powerful mandibles.

BURROWING GROUND BEETLES

While most ground beetles are wedge-shaped and adapted for pushing through light soil and leaf litter in search of prey, the subfamily Scaritinae, often called burrowing ground beetles, have taken this to an extreme. Many species have cylindrical bodies with large, pointed mandibles, strong back legs for pushing, and fossorial (adapted for digging) fore legs, allowing them to lead a mole-like subterranean way of life. The mandibles are strong, not just for boring through the soil, but also for holding onto and crunching on strong or heavily armored prey such as other beetles, snails, or large worms. Some scaritines will also kill and eat vertebrates such as frogs, lizards, or even small snakes.

Many species of burrowing ground beetles live in mud at the edges of rivers and streams, or in coastal salt marshes or mangroves near the sea, and they are

SUBFAMILY
Scaritinae

KNOWN SPECIES
1,900

DISTRIBUTION
Worldwide except Antarctica. Concentrated in the tropics and warmer regions of the southern hemisphere, particularly Africa, Madagascar, and Australia

HABITAT
Found burrowing in moist soil, in wet tropical forests, near rivers, and close to the shores of lakes or the sea. Some species can be very common in mangrove swamps or salt marshes, and others occur in gardens or agricultural settings, especially rice paddy fields. A few prefer dry, sandy environments, including semideserts

SIZE
1.5–65 mm

rarely seen, coming out at night to forage on the surface. Many species fly by night and are attracted to lights, where they may feed on moths and other insects that are also attracted. Some species of the genus *Dyschirius* invade colonies of burrowing staphylinid beetles, for example the genus *Bledius*, and eat the adults and larvae.

One of the largest burrowing ground beetles is the formidable *Ochyropus gigas* from West and Central Africa, which has very elongated predatory mandibles and resembles a stag beetle, especially while in flight. Another large African species, *Mamboicus bittencourtae*, was newly discovered in 2018 in the unique habitats of the inselberg mountains of northern Mozambique, by scientists researching amphibians.

ABOVE | *Mouhotia batesi* This colorful, large, robust species is from Indochina, and has appeared on postage stamps from both Thailand and Laos.

OPPOSITE | *Carenum speciosum* Most scaritines are brown or black, but the large Australian species shown here has bright metallic colors.

RIGHT | *Pasimachus mexicanus* This large predatory species can be found in fields and gardens as well as forests in southern Mexico.

DIET
Voracious predators as adults and larvae. Scaritines feed on anything they can overpower, including large beetles and small vertebrates

NOTES
Although most Scaritinae burrow in the ground, they hunt at night or fly in search of mates or new habitats. However, some species have lost their eyes and the ability to fly, and live entirely underground or in deep cave systems

CRAWLING WATER BEETLES

The suborder Adephaga is divided into 11 families, of which nine are mainly or entirely aquatic. One of the smaller aquatic families, which is morphologically and ecologically quite distinct from the others, is the crawling water beetle family Haliplidae, which is widely distributed but usually not particularly abundant, in still or slow-moving fresh water throughout most of the world.

Haliplidae are called crawling water beetles because they swim with a peculiar irregular stroke, where each of the pairs of swimming legs alternate, and they frequently climb over submerged vegetation instead of swimming in open water. The legs are not strongly adapted for swimming, and have not developed into paddles as in Dytiscidae or Hygrobiidae. The adults of many species disperse between water bodies by flying during the night, and are attracted to electric lights. Some species, especially those living in large lakes, tend to lose the ability to fly, but like other Adephaga (notably Carabidae) this may vary from population to population, with the same species producing flying and flightless individuals in response to different ecological conditions.

BELOW | *Haliplus ruficollis* feeding on freshwater bryozoans in Europe. Note the large air bubble held under the coxal plates.

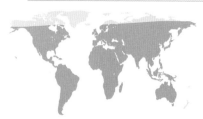

FAMILY
Haliplidae

KNOWN SPECIES
220

DISTRIBUTION
Worldwide, but mostly found in the Holarctic region

HABITAT
Slow-flowing or still water with algae. Some species are found in fast-flowing or even brackish water

SIZE
2–4 mm

DIET
Larvae primarily or entirely feed on algae. Adults eat a wider variety of food, including slow-moving, immobile, or dead invertebrates

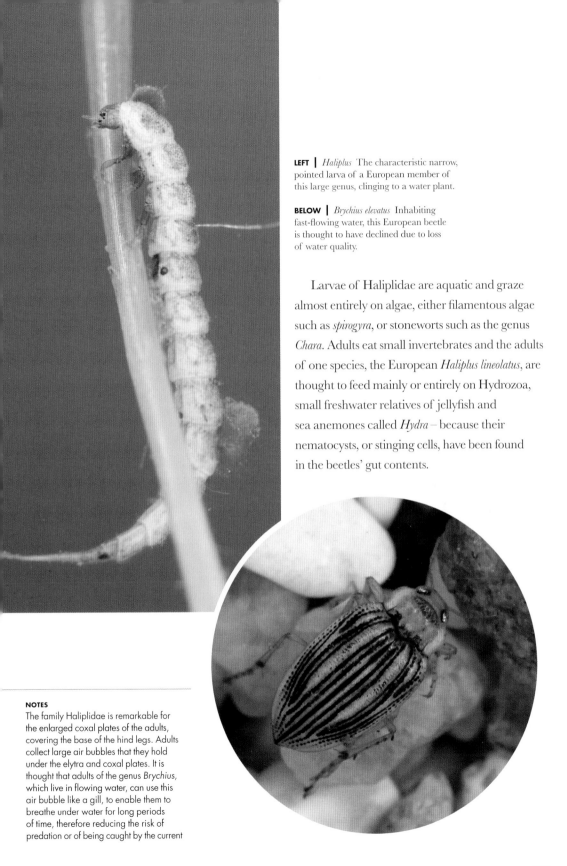

LEFT | *Haliplus* The characteristic narrow, pointed larva of a European member of this large genus, clinging to a water plant.

BELOW | *Brychius elevatus* Inhabiting fast-flowing water, this European beetle is thought to have declined due to loss of water quality.

Larvae of Haliplidae are aquatic and graze almost entirely on algae, either filamentous algae such as *spirogyra*, or stoneworts such as the genus *Chara*. Adults eat small invertebrates and the adults of one species, the European *Haliplus lineolatus*, are thought to feed mainly or entirely on Hydrozoa, small freshwater relatives of jellyfish and sea anemones called *Hydra* – because their nematocysts, or stinging cells, have been found in the beetles' gut contents.

NOTES

The family Haliplidae is remarkable for the enlarged coxal plates of the adults, covering the base of the hind legs. Adults collect large air bubbles that they hold under the elytra and coxal plates. It is thought that adults of the genus *Brychius*, which live in flowing water, can use this air bubble like a gill, to enable them to breathe under water for long periods of time, therefore reducing the risk of predation or of being caught by the current

GREAT DIVING BEETLES

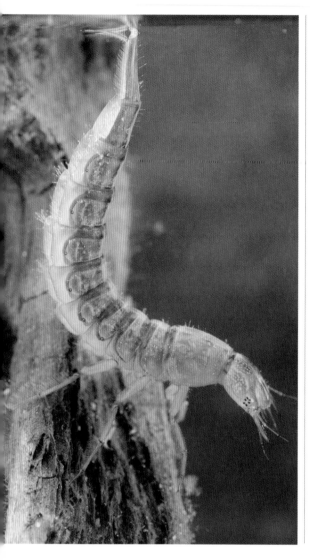

Although comprising only 10 percent of the species of diving beetles, the subfamily Dytiscinae, known as the great diving beetles, includes some of the largest and most conspicuous of all aquatic beetles. The biggest species, at almost 2 in (5 cm), is *Megadytes ducalis*; until recently it was known from only a single specimen apparently found in a water-filled canoe along the Amazon River in the late 1800s, and kept at the Natural History Museum in London.

All Dytiscinae are rapacious hunters as adults and larvae. The large, yellowish-brown, striped larvae, called "water tigers," have hollow, sickle-shaped mandibles that are used to liquify their prey with enzymes before sucking up its juices, and they often attack larger animals, even fish. On reaching full size, larvae crawl out of the water and pupate in a hole in the ground, the fresh adult either returning to the pond or dispersing in search of new water bodies. Dispersing adults fly powerfully at night, and may be attracted to lights, or drop down onto shining objects such as cars or glasshouses, mistaking the reflected moonlight for a water surface. They have even been found at oases deep in the Sahara Desert.

LEFT | *Dytiscus marginalis* The voracious larvae of this European great diving beetle often eat vertebrates such as fish, newts, and tadpoles.

SUBFAMILY
Dytiscinae

KNOWN SPECIES
380–400

DISTRIBUTION
Worldwide except Antarctica. Large species are common in northern climates as well as in the tropics

HABITAT
Fresh water, including ponds, lakes, rivers, and some brackish areas near the coast. Can be found in very small, recently formed pools

SIZE
10–45 mm

DIET
Carnivorous. Adults and larvae feed on aquatic arthropods, worms, fish, and amphibians. Can be a minor pest in fish or shrimp farms

Most great diving beetles are greenish, oily black or brown, often with yellow margins and a patterned underside. The male has suction pads on the fore legs to hold the female during mating, and the females of several genera have evolved ridges on their backs to prevent unwanted attention.

ABOVE | *Eretes australis* An Australian diving beetle with a large air bubble. The hind legs are for swimming, the others for anchoring on substrate.

RIGHT | *Dytiscus marginalis* Female (left) and male (right) adults. The female is recognizable from the ridged elytra, the male by the suctorial pads found on the fore legs.

NOTES
In many parts of the world, the adults of Dytiscidae, particularly the large-bodied genera *Cybister* and *Dytiscus*, form part of the human diet. Although crunchy in texture, they are high in protein and readily available. However, there have been concerns that because of their relatively long life and predatory habits, they may accumulate pollutants

SMALL DIVING BEETLES

Hydroporinae is the largest subfamily in the diving beetle family Dytiscidae, but includes some of the smallest diving beetles. They are distinguished from other subfamilies by the absence of a visible scutellum, characteristic segmentation of the tarsi (feet) in adults, and by larva having a long, hornlike projection in the center of the head, which, together with the mandibles, is used for catching and restraining prey. Like other Dytiscidae, adults and larvae are predators, usually of small worms or insect larvae. The aquatic larvae of Hydroporinae swim well, and crawl out of the water to pupate in the soil or sand on the bank. The adults of most species can fly, especially those that depend on temporary pools.

Hydroporinae occur throughout the world, but species are usually closely associated with specific habitat types, from temporary rain pools to large lakes. Generalist species are often among the first beetles to colonize new water bodies, such as rain-filled tire tracks or even paddling pools. In more complex environments,

BELOW | *Hydroporus palustris* This widespread Eurasian species renews its air supply at the water surface.

SUBFAMILY
Hydroporinae

KNOWN SPECIES
2,200

DISTRIBUTION
Worldwide except Antarctica

HABITAT
Ponds, pools, rivers, and streams, with a few species found in cave systems or subterranean aquifers

SIZE
1–8 mm

DIET
Adults and larvae are predators of other aquatic invertebrates, and may also scavenge on dead organic matter

NOTES
A few Hydroporinae, such as the genus *Geodessus* from India, have left the water as adults and instead live like ground beetles in leaf litter. Some other species are even more

numerous species can be found together, because of subtle differences in microhabitat.

The largest genus is *Hydroporus*, with around 200 described species and mostly marked in intricate patterns of brown, black, and yellow. Many are difficult to identify and need microscopic examination or dissection (but can have differing ecological requirements despite their morphological similarity). Like a lot of Adephaga, aquatic or terrestrial, adults produce defensive secretions, which can have a distinct honeylike smell caused by phenyl acetic acid.

ABOVE | *Nebrioporus elegans* This widespread European species of rivers and mountain streams may, in fact, represent several species.

BELOW | *Haideoporus texanus* The Edwards Aquifer Diving Beetle, an endangered blind subterranean species from Texas.

modified, adapted to life in subterranean streams and aquifers. An example is the North American *Haideoporus texanus*, which, living in perpetual darkness, has lost its eyes, pigmentation, and ability to fly. It was first described in 1976, only from San Marcos Pool of Edwards Aquifer in Texas, and is regarded as critically imperiled because of its small range, small population (estimated at fewer than 10,000 individuals), and vulnerability to water abstraction or pollution

OPPOSITE | *Sternotomis chrysopras* (Cerambycidae)
Named after the green gemstone chrysoprase, this
African longhorn beetle has dramatic mandibles,
but feeds on woody vegetation.

POLYPHAGA

The huge suborder Polyphaga, with over 340,000 named
species, accounts for almost 90 percent of known beetles,
sorted into 16 superfamilies and 156 families. Polyphaga
means "eating many things," and this is probably one of
the secrets of the group's extraordinary evolutionary success.
Members of Polyphaga occupy such a broad range of
ecological niches, from feeding on living plants, to dung,
to parasitism, to dead plant and animal matter, to predation.
Apart from their near-complete absence from marine
environments, there is almost no ecological niche that
is not used by members of Polyphaga, and they can be
found from the hottest deserts to well into the Arctic Circle.

The Polyphaga suborder seems to have arisen in the Permian
period, more than 250 million years ago, although such
ancient fossils are often fragmentary and difficult to interpret.
Reliable fossils of many Polyphagan families are known from
the Cretaceous period, but maybe because of the spread of
amber-producing trees, leading to much better preservation
of insect fossils. However, there is no question that the
Coleoptera suborder Polyphaga dominates terrestrial
ecosystems today.

HISTER BEETLES, WATER SCAVENGER BEETLES, AND RELATIVES

The superfamily Hydrophiloidea is one of the smaller superfamilies of Polyphaga, including only four families, each with fewer than 5,000 known species in the modern fauna. Two of the families are relatively much larger, and are covered separately below. These are the water scavenger beetles Hydrophilidae, a group of aquatic, or in a few cases, dung-living beetles with around 3,400 species, and the Histeridae, called hister or clown beetles, with around 4,300 species, most of which are predators found in decaying plant, fungal, or animal matter.

The other two families are much smaller, but are taxonomically interesting because although clearly related to the Histeridae, they do not have all the characters that would include them in this family. These are the Synteliidae, with seven species, and the Sphaeritidae, with five. Sphaeritidae are from

SUPERFAMILY
Hydrophiloidea

KNOWN SPECIES
7,712

DISTRIBUTION
Worldwide except Antarctica. Synteliidae are restricted to a few places in Asia and Mexico, and Sphaeritidae to the north of the northern hemisphere

HABITAT
Hydrophilidae in aquatic habitats or in dung, Histeridae in decaying plant, fungal, or animal matter

SIZE
1–55 mm

DIET
Histeridae are mainly predators, while Hydrophilidae adults are scavengers (many larvae are predators)

5–6.5 mm, and are found scattered across the northern hemisphere, including northern Europe, parts of Russia and China, and North America. They resemble histerids, usually having a greenish metallic sheen (although one of the Chinese species has an orange pattern). They appear to be attracted by tree sap in northern forests, where they feed and breed in the sap-impregnated soil around recently felled or wind-blown birch trees.

Synteliidae are even more unusual, and have a disjunct distribution typical of an ancient group that has become extinct over much of its range but survived in a few places; in this case, species of the same genus being known from Mexico, China, Japan, east India, and the Russian Far East.

OPPOSITE | *Syntelia sunwukong* (Synteliidae) Discovered in 2021, in Cretaceous Burmese amber, this ancient specimen seems to closely resemble the few Synteliidae that survive today.

BELOW | *Sphaerites glabratus* (Sphaeritidae) A rare species found in forest litter in northern Europe.

NOTES
The earliest unquestioned Hydrophiloidea fossils are from two famous late Jurassic localities, the Talbragar Fish Bed of Australia and the Solnhofen Limestone of Germany. Both these early fossils are Hydrophilidae, which is not surprising as their life in shallow water pools makes their preservation as fossils much more likely than Histeridae, which live in decaying matter, so probably decay themselves quickly after death

WATER SCAVENGER BEETLES

The Hydrophilidae, known as the water scavenger beetles, are the second-most species-rich group of aquatic beetles after the Dytiscidae, and can be found in freshwater bodies, especially weedy, slow-moving waters with muddy banks, throughout the world. In contrast to Dytiscidae, most hydrophilids are not very powerful swimmers, and many species spend much of their time clinging to water plants. They are called "water scavenger beetles" because adults of most species are detritivores feeding on dead plant matter, rather than active predators like the dytiscids. The larvae, on the other hand, often feed on other invertebrates, with several species being specialized consumers of water snails.

Adults collect air from the water surface, which they usually carry as a bubble on the underside of the abdomen. Hydrophilids can be distinguished from dytiscids by their shape, which is less streamlined and usually widest toward the back, rather than the characteristic oval shape of the diving beetles,

BELOW | *Berosus signaticollis* One of the exceptions to the "crawling" rule: from Europe, it is a fast and powerful swimmer.

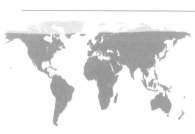

FAMILY
Hydrophilidae

KNOWN SPECIES
3,400

DISTRIBUTION
Worldwide except Antarctica

HABITAT
Many species are common in and around wetland habitats, especially in slow-moving fresh water with muddy banks, and in decay situations such as dung, compost, and fungi

SIZE
1.5–55 mm

DIET
Larvae of some genera are predators. Adults feed on a range of decaying organic matter and some vegetation

and by the palps of their mouthparts, which in hydrophilids are long, usually longer than the antennae, and are used to break the surface tension of the water when collecting air.

A quite remarkable ecological feature of Hydrophilidae is that a large proportion of species have secondarily returned to life on land. The majority of the subfamily Sphaeridiinae live in wet compost, decaying seaweed, animal dung, and other moist, nutrient-rich substrates. A few members of the genus *Helophorus* have completely escaped water and wet substrates, and larvae and adults live entirely on dry land.

ABOVE | *Hydrochara caraboides*
The European Lesser Silver Water Beetle is a typical hydrophilid, showing the air bubble carried under the body, and the enlarged palps.

BELOW | *Sphaeridium scarabaeoides*
A dung-feeding water beetle common in wet cowpats in much of Europe and North America.

NOTES
Some species are indicators of water quality. The largest European species, the 2-in (5-cm) Great Silver Water Beetle *Hydrophilus piceus*, disappeared from much of its range during the twentieth century, but is recently beginning to make a recovery in some places as pollution with agrochemicals is reduced

CLOWN OR HISTER BEETLES

The Histeridae is a medium-sized family of generally glossy black or shiny brown beetles, which are predators of larvae of other insects in a wide range of substrates. The most familiar species are associated with dung and carrion, where adults and larvae feed mainly on fly maggots. A number of species live in bird nests, or the burrows and dens of mammals, where they feed on flea larvae and other parasites, providing a useful service to their host. The same is not true of species living in ant and termite nests, which can eat the hosts themselves, or in the case of the ants, their larvae. Other species are very flattened and live under bark, preying on the larvae of other subcortical insects, while members of yet another group are cylindrical and follow wood-boring beetles into their burrows.

Despite the wide variation in shape, histerids have a distinct and instantly recognizable body plan, with angled antennae, forward-pointing mandibles, and the wing-cases or elytra shortened to reveal the last two segments of the abdomen,

RIGHT | *Sternocoelis* A myrmecophilous histerid from Morocco is tolerated by its ant host, although it eats their larvae.

FAMILY	SIZE
Histeridae	1–20 mm

KNOWN SPECIES
4,300

DIET
Adults and larvae are predators. Most species feed on the larvae of other insects

DISTRIBUTION
Worldwide except Antarctica

NOTES
The origins of the scientific name Histeridae and the English name clown beetle are both mysterious. A possible theory is that Carolus Linnaeus based the name on the Latin Histrio, an actor, and one writer has suggested that

HABITAT
Found in decaying matter such as carrion, compost, and rotting seaweed, in the nests of other animals, including mammals, birds, ants, and termites, in leaf litter or under bark

called the "pygidium." The head, antennae, and legs can be withdrawn into grooves, protecting the appendages and making the beetles look like seeds; they are so smooth and shiny that they are difficult to pick up. Almost all species can fly.

Most Histeridae are uniform black or brown, though a few have a greenish or blue metallic luster, and one or two have a pattern or red markings on the elytra.

ABOVE | *Hololepta aequalis* This flat-bodied North American beetle hunts fly larvae under tree bark. Several mites are hitching a lift on this individual.

BELOW | *Saprinus* A metallic blue histerid from Australia is associated with carrion, where it feeds on fly maggots.

this is because they "play dead." However, a species Linnaeus described, *Hister quadrimaculatus*, has striking, red, "C"-shaped markings on the elytra, which are reflected, so when viewed from the side could have reminded him of the grinning and grimacing mouths of the masks used in classical theater. "Clown" may simply be a clumsy translation of "actor"

ROVE BEETLES AND ALLIES

The vast superfamily Staphylinoidea includes more than 60,000 named species in six families, although 90 percent are in one hyperdiverse family, Staphylinidae. Families Leiodidae (3,700 species) and Hydraenidae (1,600) are fairly large, while Ptiliidae (650), Silphidae (200), and Agyrtidae (70) are comparatively small. The Ptiliidae include the tiniest known beetles, down to 0.325 mm long for *Scydosella musawasensis*, which was discovered in 1999 in Musawa, Nicaragua. They feed on fungus spores, and are almost certainly overlooked by collectors because of their minute size and often specialist habitats. The smallest beetle in Europe, 0.5 mm long but rejoicing in the name *Baranowskiella ehnstromi*, is found on only one kind of fungus, and was only

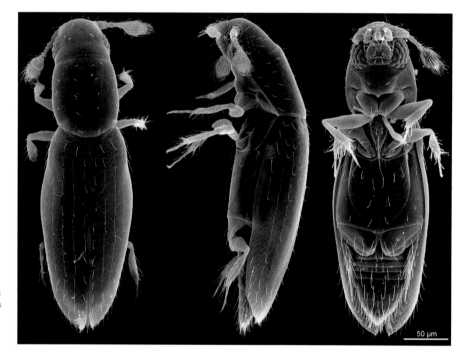

RIGHT | *Scydosella musawasensis* (Ptiliidae) At 0.325 mm long, this ptiliid may be the smallest known adult insect. It lives in fungi in South and Central American forests.

50 μm

SUPERFAMILY Staphylinoidea	**SIZE** 0.325–50 mm
KNOWN SPECIES 62,220	**DIET** Most Staphylinidae are predators, while some other families are fungivores (Ptiliidae, some Leiodidae) or carrion feeders (some Agyrtidae, Silphidae, and Leiodidae)
DISTRIBUTION Worldwide except Antarctica (but found on sub-Antarctic islands)	
HABITAT Almost all environments support some Staphylinoidea, from mountaintops to the seashore	**NOTES** Most Staphylinidae have the wing-cases (or "elytra") shortened, giving them their characteristic shape and making the body

LEFT | *Coelometopon* (Hydraenidae) These beetles, this one in South Africa, crawl on rocks in the spray-zone of waterfalls grazing algae. They resemble other water beetles but are not closely related to them.

BELOW | *Necrophilus pettiti* (Agyrtidae) This North American primitive carrion beetle might resemble Silphidae, but is actually more closely related to Leiodidae.

discovered and named in 1997, despite inhabiting an entomologically well-studied continent.

Ptiliids are called featherwing beetles because their wings are reduced to a featherlike strut with a fringe of bristles, with which they "row" through the air (which is very dense to such a small animal). Some travel greater distances by hitchhiking on birds. Ptiliid reproduction is also modified for small size, the females producing one egg at a time. For this reason, they have surprisingly long adult lives of at least several months.

Hydraenidae are aquatic, and have a strong superficial resemblance to Hydrophilidae, although they are usually smaller. They are often found in brackish water near the coast, even tide pools, where they feed on algae and organic debris. Agyrtidae are a very small family, mainly found in the north of the northern hemisphere and in New Zealand, where they scavenge on carrion and organic matter.

more flexible to enter holes and crawl through substrates. This has been proposed as a reason for their extraordinary success. The other five families of Staphylinoidea are more typically beetle-shaped, with the elytra extending to, or nearly to, the tip of the abdomen, but none of these families reach even 10 percent of the diversity and species richness of the Staphylinidae

ROUND FUNGUS AND SMALL CARRION BEETLES

The Leiodidae is the second largest family in the Staphylinoidea, even though at around 3,700 species it is tiny compared to the hyperdiverse Staphylinidae. Leiodidae is divided into several subfamilies, and a large proportion of them have the eighth segment of the adult antenna very small compared to the seventh and ninth, an immediately recognizable feature despite the considerable variation of morphology within the family.

Two main subfamilies are likely to be found. The domed, shiny subfamily Leiodinae (round fungus beetles) inhabit moist, rotting, fungal fruiting bodies, under bark, on slime molds, and similar situations, and fly in the evening, when they can be collected by sweeping a net through vegetation. Many species are only associated with subterranean fungi, and are rarely seen. The subfamily Cholevinae (small carrion beetles) are flatter and more elongated with longer legs, and are found living as scavengers in the burrows of mammals and reptiles, or in bird nests, and on and around carrion. Highly modified

Cholevinae, the tribe Leptodirini, are found in caves, and have elongated limbs and antennae, and have often lost their eyes.

Some of the most modified of all leiodids are the subfamily Platypsyllinae, which live closely associated with certain mammals, including among the fur of the animals themselves. The Beaver Beetle

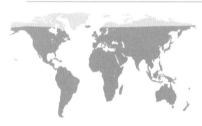

FAMILY
Leiodidae

KNOWN SPECIES
3,700

DISTRIBUTION
Worldwide except Antarctica

HABITAT
Damp places, many species in forests, others in nests and burrows of larger animals

SIZE
1–7 mm

DIET
Fungi, carrion, dead organic matter

NOTES
The genus *Glacicavicola* is one of the larger, stranger, and least known Leiodidae. The name means "ice cave dweller," and it lives in cold lava tubes, wet limestone caves, and ice caves in the Pacific Northwest of the USA, scavenging on flying insects that have been

OPPOSITE | *Zearagytodes maculifer*
Characteristic larva of this New Zealand
species, which feeds on spores of fungi.

RIGHT | *Agathidium* A typical round
fungus beetle feeds on fungi growing
on a tree stump in Pittsburgh, USA.

ABOVE | *Catops picipes*
This typical European
small carrion beetle feeds
on carrion or inhabits
mammal burrows.

Platypsyllus castoris, the only species in
its genus, is found on both the North
American and the European beaver,
which is surprising, because if
the beavers are recognized as two
distinct species, one would expect
the beetles to be too.

trapped on the ice and other organic
debris. Up to 6 mm long, with long limbs
and antennae and no eyes, it shares many
features of morphology and behavior
with cave-dwelling Cholevinae such as
Leptodirus from Europe, although it is not
currently placed in Cholevinae but in one
of the smaller subfamilies, Catopocerinae

CARRION BEETLES

The small family Silphidae, which is sometimes placed within the Staphylinidae, is divided into two major subfamilies: the Silphinae and the Nicrophorinae. The latter, comprising mainly the large genus *Nicrophorus*, are the burying beetles or sexton beetles. Male and female burying beetles, acting together, will find a small carcass such as a mouse or bird, and bury it by excavating the soil beneath it, to protect it from vertebrate scavengers. Once it is buried, they skin it, apply antibacterial substances, kill and eat any fly eggs or maggots, and the female lays eggs. Both sexes tend the growing larvae by feeding them with pieces of the carcass, which the larvae beg for like baby birds. When the larvae pupate, the adults leave in search of another carcass.

Silphinae are more generalist carrion feeders or predators, with a free-living active larva and usually lacking advanced parental care. *Necrodes* is attracted to large carcasses such as those of deer or even humans. The yellow and black *Dendroxena* are caterpillar hunters in the tree canopy, and *Phosphuga* are slug and snail predators. The distribution of most Silphidae is strongly restricted to temperate zones, especially those of the

BELOW | *Necrophila (Chrysosilpha) viridis* This beautiful diurnal carrion-feeding species from the Philippines is one of the few tropical silphids.

FAMILY
Silphidae

KNOWN SPECIES
200

DISTRIBUTION
Worldwide but concentrated in the temperate zones, particularly in the northern hemisphere. In the tropics few silphids are found, and usually at higher altitudes

HABITAT
Woodland, moorland, plains, and wherever a good supply of carrion occurs

SIZE
7–35 mm

DIET
Most species feed on carrion as larvae, and on carrion and fly maggots as adults. Some species feed on fungi, snails, or caterpillars as adults and larvae

ABOVE | *Nicrophorus americanus* The American Burying Beetle is the largest of several similar species that show advanced parental care, and is listed as critically endangered.

NOTES
The fate of one of the largest and most charismatic silphids, the American Burying Beetle, is interesting. Until the end of the 1800s it occurred widely across eastern North America but is now restricted to a few sites in Rhode Island, Oklahoma, Arkansas, and Nebraska. This catastrophic decline remains a mystery, but suggestions for the cause include light pollution, land-use change, and even the extinction of the passenger pigeon

northern hemisphere. There are more Silphidae species in North America than in South America, and more species in Britain than in Africa or Australia, which is a striking biogeographical anomaly compared to most beetles; in the tropics, the carrion-feeding niche is usually occupied by other insects, particularly members of Scarabaeoidea, and the few silphids that do occur there are generally found in the mountains.

ROVE BEETLES

Within the megadiverse beetle family Staphylinidae, often considered the largest family in the animal kingdom, the Staphylininae is the third largest subfamily, with almost 8,000 named species. Staphylininae also includes the biggest individual staphylinids, which can reach up to 2 in (5 cm) in length.

The majority of species are predators as adults and larvae, many inhabiting dung and carrion, decomposing seaweed, and other rotting materials, feeding on fly maggots which abound in such habitats. Most adult Staphylininae are able to fly, which allows them to move easily between sources of food, and they fold their wings very intricately to fit underneath the short elytra, leaving the whole abdomen free. The elongated, snake-like body with telescopic segments allows the adult beetles to be as flexible as the larvae, and to enter the holes of

SUBFAMILY
Staphylininae

KNOWN SPECIES
8,000

DISTRIBUTION
Worldwide except Antarctica

HABITAT
Found in most habitats, especially in decaying substances where prey is likely to be abundant, from beaches to tropical forest canopies

SIZE
2–50 mm

DIET
Predatory, feeding on soft-bodied prey such as fly larvae, caterpillars, worms, slugs

NOTES
One group of Staphylininae, genus *Amblyopinus* and its relatives, have adapted to live entirely in the fur and nests of small mammals, where they feed on ectoparasites. Previously, these beetles were believed to be

worms and maggots in order to feed on them with their strong mandibles.

To make up for the lack of protection from the elytra, most species have defensive glands at the apex of the abdomen that produce foul-smelling or distasteful chemicals, and when threatened they arch the body so that the apex of the abdomen is directly above the head. This stance gives them an alarming, scorpion-like appearance, as well as lining up the defensive glands with the powerful mandibles. Some species are called "cock-tail beetles" because of this posture. *Ocypus olens*, the largest species occurring in Europe, which can be seen in woods and gardens on damp evenings in late summer in search of slugs and worms, has earned the rather sinister common name "Devil's Coach Horse."

OPPOSITE | *Actinus imperialis* Another bright metallic diurnal forest species, this time from New Guinea, where it hunts caterpillars on the leaves of forest trees.

ABOVE | *Plochionocerus fulgens* A colorful diurnal species from the rainforests of South America, here folding up its flight wings under its elytra using a middle leg.

RIGHT | *Staphylinus* Unlike their somber nocturnal relative the Devil's Coach Horse, colorful European *Staphylinus* fly readily in bright sunshine.

parasitic on the mammals, feeding on their blood, but it has recently been shown that they are beneficial to their hosts, and for this reason are tolerated by the mammals and not eaten or removed during grooming. They live on mice, rats, and opossums in South and Central America and Australia

ALEOCHARINE BEETLES

The Aleocharinae is presently the largest subfamily in the Staphylinidae, possibly in the whole Coleoptera. With more than 12,000 named species, it forms about 20 percent of known rove beetle diversity worldwide. However, if the rove beetle fauna of a small, well-studied country such as Britain is examined, it is found that aleocharines comprise more than 40 percent of total rove beetle diversity. Since Aleocharinae are small and difficult to distinguish, they are usually the last beetles to be carefully studied, so it is likely that a figure of at least 40 percent can be extrapolated worldwide, implying that we are not even halfway into naming all the world's aleocharines! Taxonomic identification of aleocharines is demanding, and generally involves microscopic dissection.

In spite of the challenges of identification, this huge subfamily is of great ecological importance, and can be found almost everywhere, especially around decaying matter such as compost heaps,

BELOW | *Orphnebius* This tiny beetle, curled up on a Southeast Asian rainforest leaf, cannot be identified further than being a member of this genus.

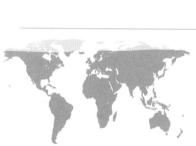

SUBFAMILY Aleocharinae	**HABITAT** Aleocharines can be found in almost every terrestrial habitat, as well as freshwater margins and sea beaches, even between the high and low tide marks. They are probably most diverse in wet tropical forests, but most of this diversity is still unknown
KNOWN SPECIES 12,000	
DISTRIBUTION Worldwide, including sub-Antarctic islands, but not yet on mainland Antarctica. The fauna of Europe and North America is best known, but most species are probably in the tropics	**SIZE** 1–12 mm
	DIET Adults are usually predators or scavengers

dead fungi, reedbeds, seaweed piles, and the nests of birds and mammals. Adults are usually predators or scavengers, but the larvae of many aleocharines are parasitoids, a way of life unusual among beetles. The larvae of the genus *Aleochara* bore into the pupae of flies; they kill the developing fly and pupate in the fly pupa, so one or more beetles hatch out instead of a fly. Another species, the small *Alaobia scapularis*, has recently been shown to develop as a parasite on the larvae and pupae of the glowworm *Lampyris noctiluca* in Britain and probably elsewhere. However, the behavior and ecology of the vast majority of the thousands of species remain unknown.

NOTES
Many aleocharines have adapted to live with ants, deceiving their aggressive hosts with chemicals that fool the ant colony into accepting them. The beetles are protected by the ants, and eat detritus as well as the ants' food supplies and even the ant larvae

ABOVE | *Aleochara bilineata* An adult beetle emerges from the pupal case of a cabbage root fly. These beetles are useful biocontrol agents against some pest flies.

BELOW | *Lomechusoides strumosus* A European species that lives in the nests of wood ants; clumps of hairs called trichomes secrete chemicals that help deceive their hosts.

ANTLIKE LITTER BEETLES

Pselaphinae have a distinctive shape, with the abdomen bulbous and shortened elytra, and the head and pronotum narrower than the abdomen. The majority of species can fly, but not readily. Even though the elytra are short, members of the subfamily do not resemble typical Staphylinidae, which is probably why Pselaphinae was until recently treated as a family in its own

right, the second largest in Staphylinoidea. The majority of species of Pselaphinae are small, about 1.5 mm being an average length. They can be collected in moss, among roots, in grass tussocks, and in the leaf litter of the forest floor, sometimes in large numbers, in both temperate forests and in the tropics, where the diversity is enormous and thousands of new species are still to be discovered.

SUBFAMILY
Pselaphinae

KNOWN SPECIES
10,000

DISTRIBUTION
Worldwide except Antarctica

HABITAT
Damp woodland and grassland, in moss and leaf litter, or in nests of ants or termites

SIZE
0.5–5.5 mm

DIET
Predators; adults and larvae eat mites and other small invertebrates

NOTES
Pselaphinae have a rich fossil history, going back to Lower Cretaceous Spanish amber. However, no association with ants has been demonstrated for any Mesozoic fossil. The earliest known myrmecophile (ant-associate)

Most Pselaphinae are free-living predators as adults and larvae, feeding on springtails, mites, and other small invertebrates. A large number of genera have become associated to some degree with ants, and the subfamily shows a whole range of ant-associations, from scavengers in the vicinity of ant nests, to true inquiline species that live in the nest and are carried and fed by their ant hosts. Some of the most highly adapted ant inquilines are in the tribe Clavigerini, which have lost their eyes; they have shortened and strengthened antennae and abdominal segments to reduce damage caused by the ants, and have trichomes, special brushlike structures that secrete liquids on which ants and their larvae can feed.

is a pselaphine, the 52 million-year-old Eocene *Protoclaviger trichodens*, which was described by one of the scientists who discovered it as "a truly transitional fossil"

ABOVE | *Claviger testaceus* From northern Europe, a highly modified eyeless pselaphine, associated with ants of the genus *Lasius*.

BELOW | *Colilodion schulzi* The only known specimen of this extraordinary species from the Philippines, named by Chinese and Swiss scientists.

SCARAB BEETLES

The Scarabaeoidea, or scarab beetles, is the smallest of the "big six" superfamilies, with almost 32,000 species spread across 2,258 genera and 12 families: Pleocomidae, Geotrupidae, Belohinidae, Passalidae, Trogidae, Glaresidae, Diphyllostomatidae, Lucanidae, Ochodaeidae, Hybosoridae, Glaphyridae, and Scarabaeidae, though as with most beetles, different experts apply different family concepts. The families vary dramatically in numbers of species: The family Scarabaeidae comprises 85 percent of the group (27,000 species), while Belohinidae are represented by a single uncommon small, brown beetle from southern Madagascar, *Belohina inexpectata* ("the unexpected one").

Adult scarabs are characterized by their lamellate antennae, the terminal segments of which form a fanlike structure of scent-detecting flaps, used for "smelling" the air to detect mates or food resources. The larvae are C-shaped, white or yellowish, fleshy grubs with clearly visible legs and a brown head, and develop buried in substrates

RIGHT | *Pleocoma dubitabilis* (Pleocomidae) A male Rain Beetle. These ancient scarabs are restricted to western North America. Males fly, often during rain, in search of buried flightless females.

OPPOSITE BELOW |
Odontotaenius disjunctus (Passalidae) The Bess Beetle or Patent Leather Beetle is the commonest passalid across much of North America. They can squeak loudly when handled.

SUPERFAMILY
Scarabaeoidea

KNOWN SPECIES
32,000

DISTRIBUTION
All continents except Antarctica

HABITAT
Most terrestrial ecosystems, especially plains and forests

SIZE
2–190 mm

DIET
Larvae usually eat decaying substrates, from wood to dung, vegetation to carrion. Some larvae attack roots. Adults may feed on dung, leaves, pollen and nectar, or not feed at all

NOTES
Varying in size and ecology, and divided into several families, adults of Scarabaeoidea are united and easily recognized by their fanlike antennae, robust stocky bodies, and fore legs with teeth often used for digging. Most species can fly, albeit slowly and noisily

such as soil, humus, or dead wood. Most species
feed primarily on living or decaying plants, though
a few (Trogidae, some Hybosoridae, some
Scarabaeidae) eat dead animal matter. A few
species are pests of agriculture or horticulture
as larvae or adults, but the vast majority of
scarabaeoids are harmless recyclers of organic
matter and play an important role in maintaining
terrestrial ecosystems.

Scarabaeoidea includes some of the biggest and
heaviest of all insects; the Goliath Beetles of Africa
(Scarabaeidae: Cetoniinae: *Goliathus*), the Atlas
Beetles of Asia (Scarabaeidae: Dynastinae:
Chalcosoma), and the Hercules and Elephant Beetles
of the neotropics (Dynastinae: *Dynastes* and
Megasoma) all inhabit tropical forest zones, and can
all exceed 4 in (10 cm) in length. Their large size,
often bright colors, and bizarre structures help
make scarabs a popular and well-studied group, but
there are also many smaller and less conspicuous
species distributed throughout the world.

EARTH-BORING DUNG BEETLES

With fewer than a thousand known species in the modern fauna, the Geotrupidae are a relatively small but widespread lineage of scarabaeoid beetles. They have a long evolutionary history, found as fossils in Early Cretaceous rocks of eastern Russia more than 130 million years old; so they were contemporaries of dinosaurs, and may even have fed on their dung before the appearance of more typical dung beetles. The name

"Earth-boring Dung Beetles" is a slight misnomer, since only one of the three major subfamilies, the Geotrupinae, is associated with dung. However, all species bore in the earth.

Geotrupinae includes the familiar Dor Beetles, or Dumbledores, that fly noisily in the evening in pastoral farming areas. These were formerly called "lousy watchmen," since their dusk flight marked the beginning of the night-watch, and they are

FAMILY
Geotrupidae

KNOWN SPECIES
920

DISTRIBUTION
Worldwide except Antarctica, particularly abundant in temperate zones

HABITAT
Geotrupinae often occur in agricultural settings as well as plains and woodlands, particularly in both northern and southern hemisphere temperate zones. Bolboceratinae are found in plains, deserts, and forests, including tropical rainforests, and are abundant in Australia. Lethrinae are sparsely distributed from the Mediterranean across Central Asia, in light woodland

SIZE
5–30 mm

OPPOSITE | *Bolbelasmus unicornis* Female (left) and male (right); a truffle-hunter, with larvae developing in subterranean fungi. It is a scarce and threatened species in Europe.

ABOVE | *Typhaeus typhoeus* A male Minotaur Beetle. Named after the mythical labyrinth-inhabiting monster, these horned beetles bury dung in tunnels up to 1.5 meters deep.

DIET
Geotrupinae feed and develop on dung, carrion, and fungi, Bolboceratinae on subterranean fungi, and Lethrinae on balls of dead leaves

NOTES
Many Geotrupidae show advanced bi-parental care, with both parents provisioning a burrow with food for the larvae. The life history of many species was unraveled by French entomologist Jean Henri Fabre in the nineteenth century

often covered with mites—although unlike true lice, these are not parasites but are phoretic, using the beetle as transportation between piles of dung. The subfamily Bolboceratinae has much more refined tastes, developing in truffles or other subterranean fungi. Many of these are uncommon and rarely encountered, though some can be a minor pest of commercial truffle farms. The third group, Lethrinae, feed on balls of dead leaves, and all three groups show advanced parental care of the developing larvae in the burrow.

Some geotrupids, particularly males, have horns or projections on the head or thorax, to defend the burrow with the larval food supply against predators or rivals.

STAG BEETLES

The stag beetle family Lucanidae is a fairly small group of large-bodied beetles, with about 1,500 species in 115 genera spread throughout the world, but most abundant and species-rich in the Asian tropics. Lucanidae get their common name "stag beetles" from the greatly enlarged mandibles of the males of some species, which resemble the antlers of a stag and serve a similar purpose, fighting between males for mates. These structures can vary greatly in size, depending on genetics but also on the amount of nutrition available to the beetle as a larva. The largest individuals, called "major males," are usually successful in competition with other males, and can pass their genes for large jaws on to the next generation. But the mandibles are also unwieldy

FAMILY
Lucanidae

KNOWN SPECIES
1,500

DISTRIBUTION
All continents except Antarctica

HABITAT
Forests and urban parklands

SIZE
5–120 mm

DIET
Larvae of almost all species feed on decaying wood, often below ground. Adults, if they eat at all, feed on sap and other sweet liquids

NOTES
Even though there are relatively few species, stag beetles are well known for their large size and the impressive mandibles of the males. The longest species, the Greater Giraffe Stag Beetle *Prosopocoilus giraffa*, is widespread throughout Asia

OPPOSITE | *Cyclommatus giraffa*
The Lesser Giraffe Stag Beetle, from the island of Borneo. Male stag beetles with elongated mandibles are abundant in the tropics.

RIGHT | *Lucanus cervus*
The classic European stag beetle. The biggest beetle in northern Europe, it has adapted to living in urban parks and even gardens.

BELOW RIGHT |
Phalacrognathus muelleri
The Rainbow Stag Beetle from tropical Australasia is one of the most colorful of all stag beetles, and is popular as a pet.

and can be an impediment, particularly in flight. The females, which are longer lived as adults, lack these elaborate ornaments and have smaller, more functional mandibles.

The larvae of most stag beetles live and feed in decaying wood, usually below ground level, where the moisture content is relatively high. Larval development, especially in temperate climates, may take several years as the wood is not very nutritious. After pupation, the active adult life of most species lasts no longer than a few weeks, and many of them take little or no food as adults.

In many countries, lucanids are among the largest and most conspicuous of beetles, and a considerable folklore and culture has developed around them. In some Asian countries they are used in gambling, and bets are placed on the outcomes of fights between males such as those of the genus *Prosopocoilus*. The group is also popular with collectors and hobbyists, both for insect collections and as exotic pets, and live stag beetles can be purchased from slot machines in Japan. Stag beetles can be a surprising sight on city streets on warm summer evenings in many parts of the world, since

several species have adapted well to life in human-modified habitats, developing in street trees and trees in urban parks and cemeteries. On the other hand, the flightless stag beetles of the genus *Colophon*, restricted to several mountaintops in South Africa, have some of the smallest natural ranges of any beetles, and are among the only beetles given legal protection under the Convention on International Trade in Endangered Species (CITES).

SCAVENGER SCARAB BEETLES

ybosoridae, sometimes called scavenger scarab beetles, is a small family of Scarabaeoidea, almost restricted to the tropics. There are two subfamilies, Hybosorinae and Ceratocanthinae, which were long thought to be separate families because their appearance and ecology is completely different. But DNA and detailed morphological study have shown them to be closely related.

Although there are few species, Hybosorinae can be found in large numbers. They earn their English name because many of them feed on dead organic matter, and they fill the carrion beetle niche in the tropics that is occupied by Silphidae and Leiodidae in temperate regions. The genera *Phaeochroops* and *Phaeochrous* are sometimes found in their hundreds under slightly dried-up, road-killed carcasses, which can seem to

FAMILY
Hybosoridae

KNOWN SPECIES
570

DISTRIBUTION
Worldwide, but almost absent from temperate latitudes and concentrated in the tropics

HABITAT
Forests, with some hybosorinae being found in grasslands

SIZE
3–15 mm

DIET
Scavengers; many Hybosorinae feed on carrion, while others live and feed in dead wood or leaf litter, as do most Ceratocanthinae. A few Ceratocanthinae, which are rare, appear to develop and feed in termite colonies

move because of the heaving mass of beetles underneath them.

Members of Ceratocanthinae are not carrion feeders, but live in the leaf litter on the forest floor. They are distinctive for their ability to roll themselves into a ball, called "conglobation." The head, thorax, and abdomen, as well as small plates on the legs, are involved in this process, the result of which is an almost perfect smooth-sided spheroid, sometimes camouflaged and sometimes glossy metallic. These spheroids are difficult for a predator such as a bird or an ant to grip or pick up. All Ceratocanthinae can roll up into balls in this way, except a few rare, poorly known Neotropical genera such as *Ivieolus* and *Scarabatermes*, which have evolved a different shape, probably from living with termites.

ABOVE | *Eusphaeropeltis* This glossy, spherical Malaysian ceratocanthine is difficult for a predator to pick up, and resembles a water droplet.

OPPOSITE | *Phaeochrous* A typical carrion-feeding Hybosorinae, from Australia.

RIGHT | *Madrasostes variolosum* This ceratocanthine from Singapore has a rough texture, and resembles a seed or animal dropping.

NOTES
Ceratocanthines conglobate, or roll up into a ball, as a defense mechanism, but it doesn't always save them from being eaten. Specimens of undigested ceratocanthines were found inside the stomach of a toad specimen in the Natural History Museum's collection in London, collected over 100 years ago. It was only by identifying the beetles that scientists could find out where the toad was from, because it was incorrectly labeled

TRUE DUNG BEETLES

The true dung beetles, or scarab dung beetles, (subfamily Scarabaeinae) have a short fossil history. The oldest unquestioned fossil is Eocene, 53 million years old, well into the Age of Mammals. An association with mammal dung remains today, and they are essential decomposers in terrestrial ecosystems, clearing up the waste products of both wild and domestic animals. Every egg laid by a true dung beetle is accompanied by a large ball of dung, enough to provide for the larva's whole development, so a great quantity of dung is removed and buried by these industrious insects. True dung beetles are divided into "rollers," which mold the dung into a ball and roll it away from competitors before burying it, and "tunnelers," which dig a burial shaft directly under the dung pile.

Especially in South America, where much of the "megafauna" (such as giant ground sloths and elephant-like gomphotheres) is extinct, many true dung beetles feed and breed on carrion. These include most of the brightly colored "rainbow scarabs" of the tribe Phanaeini. The largest scarabaeines are the genus *Heliocopris*, more than 2 in (5 cm) long, tunnelers often found around elephants and rhinoceroses, where these producers of large-enough dung piles still exist in the tropics of Africa and Asia. Most dung beetles are powerful fliers to move from one dung pile to the next, but in some parts of Africa flightless dung beetles such as the genus *Mnematium* can be found. The vast ruminant herds of the African plains allow them to walk between dung piles.

SUBFAMILY
Scarabaeinae

KNOWN SPECIES
6,000+

DISTRIBUTION
Worldwide, most diverse in tropics and subtropics, much rarer in northern and southern climates (e.g. few species in Canada, UK, Chile, New Zealand)

HABITAT
Many species live in grasslands and savannahs with big herds of grazing vertebrates. Tropical forests are rich in smaller species, and some perch on leaves in the canopy waiting for dung

SIZE
2–55 mm

DIET
Most species develop and feed on vertebrate dung. Several species,

ABOVE | *Sulcophanaeus imperator*
An impressive horned male of the
South American Emperor Dung
Beetle, a "rainbow scarab" that feeds
mainly on carrion.

OPPOSITE | *Kheper nigroaeneus* One
of the common dung beetles of the
African savannah, here with a ball
of fresh dung.

RIGHT | *Onthophagus taurus* This
European species demonstrates the
large, backward-curved horns that
are only found in the male.

especially in South America, have adopted
a carrion diet, and a few live in ant nests

NOTES
Scientists use dung beetle populations
to learn about the ecology of an area,
especially in the tropics. They are easy to
attract (the bait is freely available), relatively
easy to identify, and large diverse
populations indicate a healthy mammal
fauna. DNA from scarab guts can help
show what mammals live in an area

RHINOCEROS BEETLES

The subfamily Dynastinae, known as the rhinoceros beetles, is a medium-sized group of between 1,000 and 2,000 species. The males are often ornamented with extravagant horns, hence the name "rhinoceros." These horns serve a similar function to a peacock's tail, as they are used for mating displays and signaling, but they are also used, like the antlers of stags (or stag beetles), for fighting with other males for control of the unornamented female, which is usually in the nearby foliage awaiting the outcome of the battle. In spite of these huge appendages, almost all rhinoceros beetles are able to fly (even though they are not especially elegant!). They may fly noisily around street lamps and other light sources on hot tropical nights, but only when the temperature and humidity reach certain levels.

Rhinoceros beetles include the heaviest of all beetles (a larva of the Central American Actaeon Beetle *Megasoma actaeon* has been weighed at 7 ounces/200 g, about the same as a hamster). These massive white grubs feed in humus-rich

LEFT | *Chalcosoma moellenkampi*
A male Bornean Atlas Beetle, or Three-Horned Rhino Beetle, the largest beetle on the island of Borneo.

SUBFAMILY
Dynastinae

KNOWN SPECIES
1,500

DISTRIBUTION
Warm places, tropics, and subtropics. Rare or absent in temperate countries

HABITAT
Forests, anywhere where there is an available supply of decaying plant matter such as wood mold or compost for

development, even in and around tropical cities in parks and gardens

SIZE
1–19 cm

DIET
Larvae eat decaying plant matter such as wood mold, sawdust, compost, and humus-rich soil. Adults may feed on sweet sap or nectar but in many cases do not feed at all

compost and decayed wood. If you count the horn as well as the body itself, Dynastinae can also boast the longest adult beetle: the Hercules Beetle *Dynastes hercules* can reach 7 ½ in (19 cm)—although half of that is the horn!

Some rhinoceros beetles, such as the Palm Rhinoceros *Oryctes rhinoceros*, are pests of coconut palms in Hawaii and the Pacific Islands. However, most species inhabit tropical forest areas and do no harm, and in spite of their imposing appearance, they cannot bite or hurt humans. Many species have no horns in either sex.

ABOVE | *Dynastes hercules* A mating pair of Hercules Beetles from South America, which are among the longest of all beetles (if you include the male's horn).

RIGHT | *Hexodon* One of many species of this hornless rhinoceros beetle genus, which resemble tenebrionid beetles and live in sandy Madagascan forests.

NOTES
In parts of Asia, such as Thailand, common rhinoceros beetles, usually *Xylotrupes gideon*, are popularly used in gambling. A female is imprisoned in a bamboo cage and two males placed on it and encouraged to fight, with bets placed on which will win

SHINING LEAF CHAFERS

BELOW | *Macraspis chrysis* From Venezuela, this beetle develops as a larva in dead wood. *Macraspis* means "big shield," referring to the large scutellum.

The shining leaf chafers (Rutelinae) are closely related to the rhinoceros beetles (Dynastinae), although they tend to be more metallic and fewer species have horns in the male. Their larvae are the typical C-shaped, white grubs of Scarabaeoidea, and feed in the soil on roots, or in compost or decaying wood usually on the ground. Adults fly, feeding on leaves. Many species are nocturnal and attracted to artificial light. The tribe Anomalini includes the huge genera *Anomala* and *Mimela*, which are very species-rich, especially in tropical Asia, and can be difficult to identify, requiring dissection.

A few species are pests, for example the Japanese beetle *Popillia japonica* was introduced to North America in the early twentieth century and rapidly established, attacking the leaves and flowers, and the larvae the roots of garden plants. It has now also reached parts of Europe. However, most Rutelinae live harmlessly in the world's tropical forests, and some are much sought

SUBFAMILY
Rutelinae

KNOWN SPECIES
4,000

DISTRIBUTION
Worldwide except Antarctica, most abundant in the tropics

HABITAT
Tropical forests, plains, pastures, sand dunes

SIZE
5–60 mm

DIET
Adults feed on leaves or petals, larvae are saprophagous in soil, leaf litter, or dead wood

NOTES
The metallic coloration of adults of some shining leaf chafers is not a pigment, but a structural color caused by multilayer reflection of light by microscopic structures in the exoskeleton. This means that the color doesn't fade after death, and specimens in collections retain their metallic sheen. It is initially difficult

LEFT | *Chrysina aureola* The genus *Chrysina*, the jewel chafers, includes some of the most metallic of all beetles. This one is from Ecuador.

after by collectors for their bright metallic colors. These are most pronounced in members of the tribe Rutelinae, which includes the Neotropical jewel chafers, genus *Chrysina*, which inhabit mountain cloud forests of South and particularly Central America, and many of which have a striking gold or silver reflection. The Australian Christmas beetles, genus *Anoplognathus*, can be almost as brightly metallic, and can be found in suburban areas. They earn their name from their emergence around Christmas, at the height of the Australian summer, and possibly also for their resemblance to Christmas-tree baubles. They may be declining in some areas.

RIGHT | *Kibakoganea formosana* A species of horned leaf chafer from Taiwan. Only the males have long projections on the mandibles like this.

to imagine the advantage of being seemingly so conspicuous, for a beetle that aims to avoid predators, but in the dappled light and shade of a tropical cloud-forest canopy, or resting on moving leaves that also hold multiple droplets of water, the reflective surfaces of adult *Chrysina* can make them surprisingly difficult to see

MARSH BEETLES AND ALLIES

With just over a thousand species worldwide, Scirtoidea is one of the smallest superfamilies of Polyphaga, and is traditionally divided into four families: Decliniidae, Eucinetidae, Clambidae, and Scirtidae. Some recent studies suggest that these divisions may need to change. Eighty percent of Scirtoidea are in the family Scirtidae, or marsh beetles. Short-lived as adults, they are found on vegetation, and have a soft, fragile exoskeleton. They spend most of their life as woodlouse-shaped aquatic larvae with distinctive, exceptionally long, multi-segmented antennae. Most larvae live in ponds, pools, and marshes, in submerged leaf litter, or hide between leaves of reeds or rushes. Some live in forests, in water that accumulates in holes in trees, or in between the leaves of bromeliads or orchids in the rainforest canopy. In Europe the uncommon genus *Prionocyphon* is found as larvae in water-filled rot holes in beech trees, and larvae are easier to find than the short-lived adults, which are evident only for a few weeks or even days in summer.

The Clambidae is a family of fewer than 200 round, black or brown beetles that live in decaying vegetation. Although probably quite common, and distributed worldwide, they are tiny and rarely seen, and easily mistaken for mites. Adults can roll into

BELOW | *Scirtidae* A typical woodlouse-like larva grazing on moss and algae in still water.

SUPERFAMILY
Scirtoidea

KNOWN SPECIES
1,025

DISTRIBUTION
Worldwide, but almost equally abundant in tropical and temperate environments

HABITAT
Associated with wetlands. Many species use water trapped in plants in the forest canopy, particularly in wet tropical forests

SIZE
1–10 mm

DIET
Adults and larvae graze on algae, moss, fungi, and decaying vegetation. Eucinetidae are associated with fungus-like slime molds

NOTES
Recent DNA studies indicate that Scirtoidea are among the most ancient groups of Polyphaga. However, because they are small and fragile as larvae and adults, they do not

a ball, increasing this resemblance. Eucinetidae
includes around 50 usually rare species; they are
able to jump. The last family, Decliniidae, from
the far east of Asia, was discovered in the 1990s;
there are only two known species, of which
little is known.

preserve easily as fossils, so their known
fossil record goes back only as far as
the Cretaceous period. Supposed fossil
Clambidae are known from early
Cretaceous Lebanese amber, and Scirtidae
from mid-Cretaceous Burmese amber. It
is likely that Scirtoidea date back to much
earlier, but fossil evidence for this is still
to be discovered

SOFT-BODIED PLANT AND CICADA PARASITE BEETLES

BELOW | *Rhipicera* (Rhipiceridae)
This Australian cicada parasite has
extraordinary elaborate antennae for
detecting scent particles in the air.

A nother of the smaller superfamilies of
Polyphaga, Dascilloidea has only two families
and a total of about 150 species. Usually found in
small numbers around the world, some species can
occasionally be abundant. Adults are short-lived,
and so not often seen, and in a large proportion
of species females are unable to fly, which has
no doubt impeded their ability to disperse.
Dascilloidea is considered closely related to
Byrrhoidea, and several families have changed
from one superfamily to the other.

The two families presently placed in Dascilloidea
are the soft-bodied plant beetles (Dascillidae), which
have about 15 genera and 80 species, and the cicada
parasite beetles (Rhipiceridae), with 7 genera and
70 species. Dascillidae have soil-living, root-feeding
larvae, while in Rhipiceridae, as their common
name suggests, larvae attack subterranean nymphs
of cicadas (Hemiptera: Cicadidae), feeding as an
external parasite.

The common European *Dascillus cervinus* is
called the Orchid Beetle because the adults are
often found in well-drained, flower-rich meadows
where wild orchids grow, but although the larvae
are soil dwelling and root feeding, they may
only incidentally feed on orchid roots, as they

SUPERFAMILY
Dascilloidea

KNOWN SPECIES
150

DISTRIBUTION
Worldwide except Antarctica, but few
species and rarely seen

HABITAT
From tropical forest to semidesert

SIZE
10–20 mm

DIET
Larvae of Dascillidae feed on plant roots,
while those of Rhipiceridae are parasites
on the nymphs of cicadas

NOTES
Males of the genus *Rhipicera* have
extraordinary plumose antennae, which are
designed to maximize their surface area to
make them more efficient at detecting tiny
quantities of chemicals, in this case the
female pheromone, in the air. The males
climb to the top of a branch or grass stem

often occur in places where orchids do not.

Some Dascillidae, like the North American *Anorus* and north African *Emmita*, live in desert and semidesert environments. Flightless females remain in the larval burrow and produce pheromones to attract flying males. In some genera, females are still unknown, but these are assumed to be flightless and subterranean.

ABOVE | *Dascillus davidsoni* (Dascillidae) A fully winged male from western North America. In contrast, the female has full elytra but shortened flight wings.

BELOW | *Sandalus niger* (Rhipiceridae) A male of the Cedar Beetle, a widespread cicada parasite in eastern USA and southeast Canada.

and unfurl their antennae, reading the wind to find out where the nearest females are. If humans could imitate the scent-detecting apparatus or these beetles, we could design more effective equipment to detect minute quantities of explosives or toxins

FLAT-HEADED JEWEL BEETLES

The jewel beetles, also called metallic wood-boring beetles (family Buprestidae), number more than 14,500 species worldwide. Almost all buprestids have short-lived, brightly colored, and actively flying adults, and larvae that develop inside wood, usually on living trees, leaving sinuous burrows beneath the bark. Larvae of some genera mine in leaves, feeding on the living tissue between the cuticles, and leaving behind a characteristic serpentine-shaped tunnel of dead leaf tissue stuffed with droppings. A few large species have larvae that develop in the soil, feeding on plant roots. Most buprestids are found in the tropics and subtropics, and only a few species extend into northern and southern latitudes.

The subfamily Agrilinae, called flat-headed jewel beetles, is one of the largest subfamilies, and very widespread and taxonomically diverse. It includes most of the leaf miners, genera such as *Trachys*, *Brachys*, and *Habroloma*. Among the living-wood feeders is the genus *Agrilus*, which has over 2,900 named species, making it a contender to be the biggest genus in the animal kingdom. *Agrilus* are small, iridescent, bullet-shaped beetles, and the

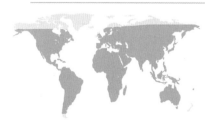

SUBFAMILY
Agrilinae

KNOWN SPECIES
7,200

DISTRIBUTION
Worldwide, most abundant in the tropics and subtropics. Many species in tropical Asia, Africa, northern Australia, and South and Central America, with fewer in Europe and North America

HABITAT
Wherever host plants grow, especially warm areas. Some species are found in arid habitats, others in rainforests

SIZE
2–22 mm

DIET
Larvae are either wood or stem feeders, or leaf miners. Adults of many species hardly eat at all during their short adult life, but may nibble leaves or flowers

short-lived, fast-moving adults are not often seen. The usual sign of their presence is the D-shaped exit holes the adults leave in the bark of living trees when they emerge in the spring and summer, and heavy infestations may kill the tree. Like many species-rich, plant-feeding beetles, *Agrilus* are host-specific, using only a single genus or species of plant. Some species attack woody shrubs instead of trees.

ABOVE | *Habroloma lepidopterum*
A leaf miner from Singapore, the related *H. myrmecophila* mines in leaves used to build weaver ant nests.

OPPOSITE | *Agrilus australasiae* This Acacia Flat-Headed Jewel Beetle from Australia is a typical member of the huge genus *Agrilus*.

RIGHT | *Agrilus ruficollis* This Red Necked Cane Borer makes swellings in the stalks of blackberry and related plants in North America.

NOTES
Some Agrilinae are extending their geographical range as a result of climate change or human movement of timber. The Emerald Ash Borer *Agrilus planipennis* from China suddenly became a serious pest of ornamental ash trees in North America, and has also reached Moscow, Russia. The list of *Agrilus* species found outdoors in Britain has doubled from 5 to 10 species in a few decades

TYPICAL JEWEL BEETLES

Most Buprestinae are fairly large, metallic, bullet-shaped diurnal species that fly rapidly in sunshine. Unlike Agrilinae, adults of several Buprestinae genera such as the Old World *Anthaxia* and the Australian *Castiarina* and *Melobasis* feed regularly, and are often found on flowers eating nectar and pollen. Some other genera, such as *Buprestis* and *Chrysobothris,* are more often seen on the wood of host trees, where they are well camouflaged despite their shining appearance. The antennae of all jewel beetles are short but have powerful scent detectors, and beetles can locate host trees from a considerable distance. Their eyesight is also excellent, and, especially when they are warmed by the midday sun, they are difficult to photograph, catch, or even approach, as they take flight at the least disturbance. Some species use sight to detect a mate: in the large, brown Australian *Julodimorpha bakewelli*, the males are attracted to discarded beer bottles that have a dimpled pattern, which apparently resembles a giant female beetle!

Some Buprestinae have very specific requirements. *Melanophila acuminata* is completely black, and lays its eggs in blackened pine trees that were recently burned. They fly very high, using special infrared sensors to detect forest fires, then descend to oviposit in the freshly charred timber.

BELOW | *Anthaxia candens* One of the most striking European beetles, which develops in cherry trees. This specimen was photographed in Hungary.

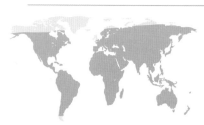

SUBFAMILY	**SIZE**
Buprestinae	5–50 mm
KNOWN SPECIES	**DIET**
3,300	Larvae of most species eat wood. Adults, if they feed at all, eat nectar and pollen or sap and small leaves
DISTRIBUTION	
Worldwide, particularly in the tropics and Australia	**NOTES**
HABITAT	The name *Buprestis*, on which Buprestinae and Buprestidae are based, means "cow sweller" or "cow burner" in ancient Greek, which is surprising for a harmless beetle!
From dry semideserts to tropical rainforests. Most species are associated with trees	

They are a good example of a beetle that follows a rare resource over huge distances, and they are so mobile that females laying eggs in burned trees in Britain may themselves have hatched in France, Spain, Scandinavia, or even farther afield.

ABOVE | *Chrysobothris caddo* perfectly camouflaged against the trunk of a fallen, dead hackberry tree in North America.

RIGHT | *Belionota tricolor* from Indonesia looks like it has been dipped in molten metal. The varying color breaks up its outline.

Pliny the Elder used *Buprestis* 2,000 years ago for an insect that kills cattle when consumed, but this should refer to blister beetles of the family Meloidae (some Melyridae also do this but they come from the Americas). It seems that Pliny's name was incorrectly applied in subsequent millennia, which shows the importance of keeping specimens in museums to link a name to the species it refers to!

LARGE JEWEL BEETLES

The subfamily Chrysochroinae is a mainly tropical group, with exceptional diversity in tropical Africa, where the genera *Sphenoptera* and *Iridotaenia* are abundant, and also in Madagascar, where there are many species of the strikingly colored genus *Polybothris*. Only a few species reach southern Europe and temperate North America. Chrysochroinae are diurnal with wood-feeding larvae. They are the third largest subfamily of Buprestidae in terms of the numbers of species, but also include some of the biggest individual species. Of the three genera of Buprestidae worldwide with species that can reach lengths of 2¾ in (7 cm), two are in Chrysochroinae (the other being the rarely seen Central Asian genus *Aaata* in the Julodinae).

SUBFAMILY
Chrysochroinae

KNOWN SPECIES
2,760

DISTRIBUTION
All continents, a strong tropical association

HABITAT
Anywhere where there are trees, so forests in particular. Can exist close to human habitation in city parks, gardens, and even street trees

SIZE
6–75 mm

DIET
Larvae of most species develop in wood. Adults feed on leaves and sap, and in some cases may visit flowers

NOTES
For centuries, the bright, striking colors and large size of buprestids, and Chrysochroinae in particular, have impressed people, especially since the metallic structural colors

Probably the largest jewel beetle of all is the Giant Ceiba Borer *Euchroma gigantea* from tropical South and Central America and the Caribbean, which has been reported at over 3 in (8 cm). For such a large species, *Euchroma gigantea* is surprisingly common and widespread, being found frequently in towns and cities in gardens or breeding in street trees. Specimens have even been found washed up in the surf on the famous Copacabana Beach in Rio de Janeiro, having flown into the sea. Although called Ceiba Borer after a genus of tropical trees, *Euchroma* develops in a wide range of wood. Larvae can be up to 6 in (15 cm) in length, and adults can be seen sunning themselves on logs. The largest Asian genus of Chrysochroinae is *Megaloxantha*, and this includes several species of metallic green and yellow beetles—although almost as long as *Euchroma*, they are more delicate.

(caused by the reflection of light rather than pigment) do not fade easily. Along with the similar genus *Sternocera* (Julodinae), members of the genera *Chrysochroa* and *Megaloxantha* have been used in Asian art, decoration, and costume design, and there are examples of Indian dresses decorated with elaborate patterns made from buprestid elytra

The Byrrhoidea, which is sometimes divided into two superfamilies, Byrrhoidea and Dryopoidea, is a varied group of 13 families of Polyphaga (with the number of described species as follows): Byrrhidae (430), Elmidae (1,500), Dryopidae (300), Lutrochidae (11), Limnichidae (390), Heteroceridae (300), Psephenidae (290), Cneoglossidae (10), Ptilodactylidae (500), Podabrocephalidae (1), Chelonariidae (250), Eulichadidae (30), and Callirhipidae (150).

The largest family, Elmidae (riffle beetles), are aquatic as both adults and larvae, with many species found in fast-flowing, well-oxygenated water; some species are indicators of water quality. Adults emerge from the water on warm nights to fly in search of new habitats, and are attracted to lights. The same is true for other aquatic

BELOW | *Byrrhus pilula* (Byrrhidae) A European Pill Beetle is able to draw its legs and antennae into special grooves to protect it from predators.

SUPERFAMILY
Byrrhoidea

KNOWN SPECIES
4,200

DISTRIBUTION
Worldwide except Antarctica

HABITAT
Varied, from mountain streams to the canopy of tropical forests. Most species avoid dry places

SIZE
1–32 mm

DIET
Varied, larvae and adults of many aquatic species graze on algae and water plants. Terrestrial larvae often in leaf litter, decaying vegetation, or dead wood

NOTES
Some Byrrhoidea species are known from huge numbers of specimens over vast areas; others from only one or two specimens.

Byrrhoidea such as Heteroceridae, Limnichidae, and Dryopidae, which between them may comprise a large proportion of a light-trap catch in the wet tropics. The adult dispersal flights makes these beetles very mobile; some Heteroceridae species can be found over the length of Africa.

The largest Byrrhoidea are Eulichadidae (forest stream beetles) and Callirhipidae (cedar beetles). Eulichadidae have aquatic larvae and terrestrial adults that resemble a large click beetle. All species except one belong to the genus *Eulichas* and are found in tropical Asia, but *Stenocolus scutellaris* is

ABOVE | *Callirhipis* (Callirhipidae) From Singapore, this beetle shows the scent-detecting flaps of the male antennae, which can detect females over great distances.

found in California, USA, which is a strange biogeographical distribution. The Callirhipidae, which have hugely developed antennal scent-detecting flaps in the male, are more widespread, and their larvae develop in dead wood. *Callirhipis philiberti*, from the Seychelles, has been reported alive in driftwood on the coast of Kenya, after traveling 1,000 miles (1,600 km) by sea.

RIGHT | *Sostea* (Dryopidae) From tropical Southeast Asia. Members of this genus are among the only long-toed water beetles with a colored metallic sheen.

The riffle beetle *Peloriolus brunneus* (Elmidae) is known from only two specimens, apparently collected by a young Charles Darwin in 1836 on the return voyage of HMS *Beagle*, on the Atlantic island of St Helena. The species has never been collected again, despite extensive surveying of St Helena, leading to discussion of whether it is extinct or whether it was mislabeled. It was possibly collected in South Africa, the *Beagle*'s previous stop, but it has never been found again there either

The superfamily Elateroidea consists of around 13 families and almost 25,000 species, although more than 90 percent of these are clustered into just four large families: Elateridae, Cantharidae, Lycidae, and Lampyridae, which appear in other sections of this book. The nine smaller families, with their number of named species, are Eucnemidae (1,500), Phengodidae (250), Throscidae (150), Artematopodidae (45), Omethidae (43), Rhagophthalmidae (30), Cerophytidae (21), Brachypsectridae (5), and Rhinorhipidae (1). There has been some recent taxonomic change, with several small families having been absorbed into larger ones following new information from DNA studies: for example, the former family Drilidae, an interesting group of snail predators, was recently transferred into the click beetle family Elateridae.

Some members of Elateroidea are unique among beetles, almost among insects, for being bioluminescent, that is, able to produce visible light, which is used in nocturnal beetles for mate attraction and sometimes for defense (when larvae or adults suddenly light up, to alarm a potential predator). The Lampyridae, the glowworm and firefly family, are the most famous light-producing beetles, but there are bioluminescent members of at least four families, the others being Elateridae, Phengodidae, and Rhagophthalmidae. Phylogenetic study of these beetles shows that the production of light must have evolved, or been lost, multiple times. A new genus *Sinopyrophorus*, just discovered in 2019 in western Yunnan, China, is also bioluminescent, and may be sufficiently different from other Elateroidea to merit the description of a new family. Major new discoveries are still being made in Coleoptera.

LEFT | *Brachypsectra fulva*
(Brachypsectridae) Larva of the Texas beetle. Short-lived adults are rarely seen, but the distinct predatory larvae are found under tree bark.

SUPERFAMILY
Elateroidea

KNOWN SPECIES
25,000

DISTRIBUTION
Worldwide, especially in warmer climates

HABITAT
Found almost everywhere, from parks and gardens, to northern tundra, to the canopy of tropical rainforests

SIZE
4–80 mm

DIET
A very wide range of diets. Many larvae are predators, though some feed on fungi or dead wood. Adults in many groups do not feed at all, though others are predatory or feed on leaves or flowers

NOTES
The Elateroidea itself was originally treated as two superfamilies, the hard-bodied

ABOVE | *Phengodes* (Phengodidae)
A male North American glowworm with scent-detecting antennae. The wormlike females have multiple luminescent spots and are called "railroad worms."

RIGHT | *Galbites* (Eucnemidae)
Eucnemidae, such as this one from Thailand, usually develop in dead wood, and closely resemble Elateridae (although most cannot click).

Elateroidea and soft-bodied Cantharoidea, but it was recently recognized that this split was not a natural one, and they were combined. In recent years, the number of families has also changed, with new techniques providing new evidence. The science of taxonomy is being refined constantly, so that classification more closely reflects the evolutionary relationships between organisms

TYPICAL CLICK BEETLES

The click beetles (family Elateridae) include at least 10,000 species found worldwide, and are now divided into 17 subfamilies. The name comes from the ability of adults to leap into the air with an audible click, caused by sliding a peg into a smooth pit on the underside of the thorax, which catapults the insect upward, helping it to right itself if it has fallen on its back, or to escape predators. Studies have shown that a clicking beetle can leap up to 12 in (30 cm) and may experience acceleration of up to 380 G. After clicking, they pull their legs and

antennae in and lie still for a time, and a predator does not know where they have landed. Many other common names, such as "snappers" and "skipjacks," are linked to this behavior.

One of the largest subfamilies is Elaterinae (typical click beetles), with around 3,500 named species. These are medium-sized beetles found in a range of habitats, and in the northern hemisphere include many species of conservation importance, as well as some pests of agriculture. Adults of the genus *Ampedus* usually have bright red or orange

SUBFAMILY
Elaterinae

KNOWN SPECIES
3,500

DISTRIBUTION
Worldwide, with many species in temperate latitudes as well as the tropics

HABITAT
From ancient forests to agricultural fields

SIZE
4–25 mm

DIET
Larvae usually predators or root feeders, while adults may graze on vegetation or feed on flowers

NOTES
Many spectacular genera of Elaterinae, such as *Elater, Ischnodes, Ampedus, Brachygonus,* and *Megapenthes,* require a dead-wood habitat to breed. As with many beetles, they rely on a supply of fallen trees. They can use small patches of suitable habitat such as old city parks and gardens, as long as some of

elytra, and predatory larvae that inhabit dead wood. They are often rare, and their presence may indicate good-quality ancient woodland. On the other hand, the genus *Agriotes* has larvae called wireworms and live in the soil, where they can be a pest, especially of root vegetables. A "worm" in a potato is quite likely the larva of a click beetle.

ABOVE | *Elater ferrugineus* One of the largest European click beetles, which is a widespread but rare species that breeds in old hollow trees.

OPPOSITE | *Cidnopus aeruginosus* A European species with short-lived diurnal flying adults, and larvae that feed among roots in the soil.

RIGHT | *Ampedus quadrisignatus* This strikingly colored European species depends on dead wood in ancient forests to survive.

the dead timber is left, but over-zealous tidying up and burning or clearing of dead trunks and branches can threaten their survival. Recently, managers of many parks and recreational forests have begun to recognize the importance of dead wood as an essential habitat for many organisms, including beetles

AGRYPNINE CLICK BEETLES

The subfamily Agrypninae is diverse and includes the largest of all click beetles. The Giant Acacia Click Beetle *Tetralobus flabellicornis* has a larva that develops inside termite mounds in tropical Africa, while the brown adult feeds on the young leaves of acacia trees, and can be 3 in (8 cm) long. In some parts of Central Africa, adults are collected for human consumption.

Another agrypnie click beetle that develops in termite mounds is *Pyrearinus termitilluminans*, one of the bioluminescent species, with a name meaning "termite illuminator." Adults and larvae glow with a bright greenish radiance, and a termite mound by night, studded with dozens of glowing larvae, is one of the wonders of the Brazilian Cerrado savanna habitat. The larvae are, in fact, predators, feeding

SUBFAMILY
Agrypninae

KNOWN SPECIES
2,300

DISTRIBUTION
Worldwide except Antarctica

HABITAT
Plains, savannas, and grasslands to tropical forests. Many species develop in dead wood, soil, or, in some cases, termite mounds

SIZE
4–80 mm

DIET
Larvae almost all predators, with large jaws, but a few eat roots. Adults in most cases feed on leaves or other vegetation, or scavenge on vegetable- and animal-based organic matter

NOTES
One of the strangest Agrypninae is the genus *Drilus*, found in the Palearctic region, and until recently placed in a separate

on the termites and flying insects attracted by their light display. *Pyrearinus* is one of a number of genera of bioluminescent click beetles, which have some of the brightest lights of any insects. The widespread New World genus *Pyrophorus* has two "headlamps" on the back of the pronotum, and a bright abdominal light, which is only visible when the elytra are open and the insect is in flight. They also light up if disturbed or threatened during the day, presumably to alarm potential predators.

Many temperate-zone Agrypninae are less impressive, but still include some large species of the genera *Alaus* and *Cryptalaus*, the larvae of which are voracious predators in dead wood. Other genera such as the European *Agrypnus* have predatory larvae that live in well-drained soil.

family, Drilidae. The bristly larva is a snail predator, entering the shell of a large snail and eating it from inside. The wingless, larvalike female also feeds on snails, while the small, soft-bodied male, looking superficially more like cantharid than a click beetle, flies in search of a mate and does not feed. This bizarre group was only placed taxonomically by using DNA

NET-WINGED BEETLES

The Lycidae are medium-sized, brightly colored, and usually slow-moving beetles, often found on trees and feeding from flowers in forested areas. While they would appear to be vulnerable, they are protected from predators by their warning coloration, which indicates that they are distasteful or toxic. When the net-like veins on their soft elytra are bent or broken, it triggers the release of a range of chemicals that often cause the predator to reject the lycid, and thereafter to avoid similarly colored insects. These well-defended beetles have become models for other insects that mimic their bright warning coloration and so gain protection, even though in many cases they lack the defensive secretions themselves. Different lycid genera adopt different patterns of warning coloration, usually involving red, yellow, metallic blue, and black.

FAMILY
Lycidae

KNOWN SPECIES
4,600

DISTRIBUTION
Worldwide except Antarctica, most abundant and species-rich in the tropics

HABITAT
Forest habitats, from light woodlands to tropical rainforests

SIZE
2–40 mm

DIET
Larvae are predatory, feeding on snails, worms, insect larvae, as well as other invertebrates. Many adults feed on nectar, while others retain the larval diet, and some do not feed at all for their short adult life

The larvae of Lycidae are flattened predators, often living under bark or in leaf litter, and feeding on snails and other soft-bodied invertebrates, while adults of most genera are short-lived and weak-flying, feeding on nectar or, in some species, not feeding at all. In some genera, such as the large Southeast Asian *Platerodrilus* (formerly known under the name *Duliticola*), females remain "larviform" for all their comparatively long life, crawling through the leaf litter of tropical forests and producing a pheromone to attract the tiny flying males. It can be difficult to tell the larviform females apart from actual larvae, since they go through no true pupation or metamorphosis.

NOTES
Lycids are abundant in warm tropical environments, but in northern latitudes they can be uncommon. In northern Europe many species are rare, and their presence can be an indication of good-quality ancient woodland

GLOWWORMS, FIREFLIES, LIGHTNING BUGS

The Lampyridae is one of the best known and most popular families of Coleoptera. Known as glowworms, fireflies, or lightning bugs, they produce a conspicuous glow during spring and summer evenings. The light is produced by chemical bioluminescence, by combining the enzyme luciferase with the compound luciferin in a specialized light emitting organ toward the tip of the abdomen. It is a cold light, lacking infrared and ultraviolet, and may appear as yellow, green, or pale red, varying between species.

Light production probably evolved as a warning to predators that the beetles were distasteful to eat, and it is still used in this way by lampyrid larvae. However, it has since been adapted as a mating signal. In the simplest cases, for example the European glowworm *Lampyris noctiluca*, the female is wingless and produces a light that attracts the

FAMILY
Lampyridae

KNOWN SPECIES
2,200

DISTRIBUTION
Worldwide, but particularly common in warmer regions

HABITAT
Forests to grasslands

SIZE
4–18 mm

DIET
Most Lampyridae are predators, as adults and larvae. Many species feed on snails

NOTES
Fireflies and their relatives have inspired poetry and art for centuries, and mating aggregations of thousands of fireflies signaling together at particular times of year are still popular attractions, especially in Asia and North America. However, these insects are declining in many parts of the world because of environmental change,

ABOVE | *Dryptelytra calocera*
From Ecuador, this species'
antennal scent-detecting flaps
suggest that it uses scent rather
than light to detect a mate.

OPPOSITE | *Lamprigera* The
wingless larviform female of this
Asian species produces a light that
attracts the smaller flying male.

RIGHT | *Aspisoma physonotum*
From South America, this beetle
is at rest camouflaged on a
rainforest leaf. *Aspisoma* means
"shield body."

flying males, which do not have a light of their own
but have large eyes to detect females. In more
complex environments, such as forests where
numerous lampyrid species coexist, individual

deforestation, and pesticide use. They are
also at risk from light pollution, where
electric lights at night can drown out the
weaker signals of the females, or attract
the males away from their habitats

species have different light signals to ensure that
they attract a mate of the correct species, and in
many cases males and females may both produce
light and use it to signal to one another.

In some cases, this mate signaling has been
subverted. For example, females of the genus
Photuris, known as the "femme fatale firefly,"
produce a light signal which resembles that of the
smaller genus *Photinus*, attracting *Photinus* males
that are seeking a mate and then eating them.

SOLDIER BEETLES

The medium-sized family Cantharidae are called soldier beetles because the bright colors of the adults of many common European species were thought to resemble historical military uniforms. Sometimes the name "sailor beetles" is used for species with elytra that are colored blue or black rather than red or orange. The whole exoskeleton of these beetles, including the elytra, is weakly sclerotized, giving the adults a soft, floppy appearance and earning them other descriptive names such as "leatherwings" and "squishy beetles."

In temperate regions, adult beetles have a short life of a few weeks, when they are easily spotted during the day among long grass, rough vegetation, or on flowers, mating or feeding on insects, pollen, and nectar. They each have a distinct adult season, with a succession of species appearing through spring and summer. The rest of the year is spent as a larva. Larvae are covered in water-repellent setae, giving them a velvety appearance, and can be abundant predators in soil, leaf litter, and among roots. Larvae of some genera such as *Malthodes* inhabit rotting wood, including dead twigs on living trees.

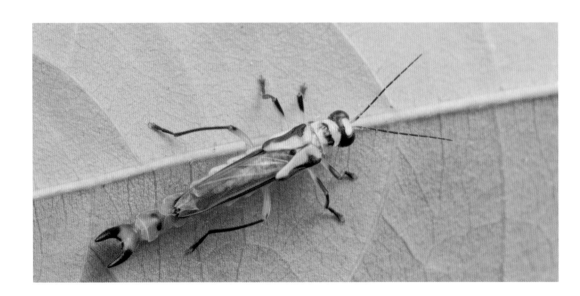

FAMILY
Cantharidae

KNOWN SPECIES
5,100

DISTRIBUTION
Worldwide except Antarctica. Most easily found, in season, in temperate climates

HABITAT
Vegetation in fields and plains, marshes, lakeside reedbeds, and forests. Some species live under stones in high mountains near the snow line

SIZE
2–30 mm

DIET
Pollen, nectar, and small insects. Many cantharids feed openly on flowers, eating nectar, pollen, and parts of the flower itself, as well as other pollinating insects

Most adult Cantharidae fly readily, but in the mountains and some islands there are species that have lost their elytra and wings; some live under stones at high altitude near the snow line.

Cantharidae were placed, with other soft-bodied beetle groups such as Lampyridae (glowworms and fireflies) and Lycidae (net-winged beetles), in their own superfamily, Cantharoidea. However, all these families are now in the superfamily Elateroidea,

with the click beetles. The earliest known fossil cantharid is *Molliberus albae* from Early Cretaceous Spanish amber, 110 million years old.

ABOVE | *Chauliognathus*
A large and widespread cantharid genus. This bicolored species from an Ecuadorian rainforest resembles *Chauliognathus domitus*.

BELOW | *Malthinus flaveolus* A small, delicate cantharid that is found in deciduous forests in May in most of Europe, in this case in Poland.

OPPOSITE | *Ichthyurus*
The forked fishtail of this common tropical genus is more pronounced in the male.

NOTES
The name Cantharidae comes from the ancient Greek *kántharos*, meaning "a beetle." The toxin cantharidin, once a famous poison and supposed aphrodisiac, is not from cantharids, but is secreted by other beetles, particularly the Spanish Fly *Lytta vesicatoria* (Meloidae), which is superficially similar, being straight-sided and soft-bodied, and had been classified with cantharids in the 1700s

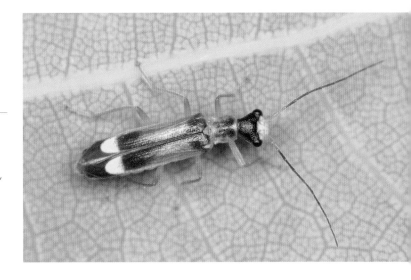

165

TOOTH-NECKED FUNGUS BEETLES AND RELATIVES

Derodontoidea is a small superfamily of Polyphaga, with an estimated 100 to 150 described species. Derodontoidea is here divided into three families: Nosodendridae (diseased-tree beetles), with two genera and 50 to 100 species; Jacobsoniidae (Jacobson's beetles), with three genera and 20 species; and Derodontidae (tooth-necked fungus beetles), with four genera and 30 species. The classification is unstable and has changed repeatedly in recent decades, with some studies placing the mysterious Jacobsoniidae in Staphylinoidea.

The family Derodontidae is most diverse in the temperate regions of the northern hemisphere, but a few species are also found in temperate South America (Chile) and New Zealand. Most species are thought to feed on fungi as adults and larvae, and unusually for beetles, are active in the colder months of the year. Members of the genus *Laricobius* are predators of aphids and their relatives— *Laricobius erichsoni* was deliberately introduced into North America as a biological control agent for balsam woolly adelgid, a serious pest of fir plantations.

LEFT | *Sarothrias cretaceus* (Jacobsoniidae) One of several such fossils from Cretaceous Burmese amber. For a rare group, Derodontoidea have a good fossil record.

SUPERFAMILY
Derodontoidea

KNOWN SPECIES
150

DISTRIBUTION
Sporadically distributed around the world, but never common

HABITAT
Derodontidae and Nosodendridae are associated with forests, while Jacobsoniidae are often found in caves

SIZE
0.65–9 mm

DIET
Mostly scavengers, eating dead organic matter. Most Derodontidae are predators, while Nosodendridae feed on sap leaking from injured trees

NOTES
Derodontoidea, especially the tiny Jacobsoniidae, are probably more abundant than we realize. Most known Jacobsoniidae

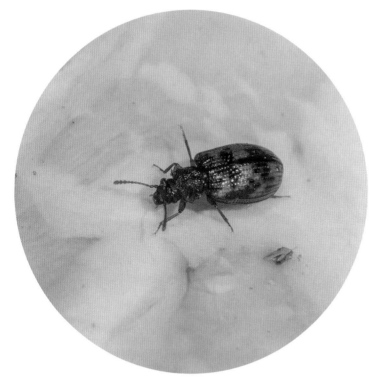

LEFT | *Derodontus maculatus*
(Derodontidae) This photograph,
showing the "toothed neck," was taken
in Maryland, USA, in December, and
reveals the beetle's winter activity.

BELOW | *Nosodendron fasciculare*
(Nosodendridae) A rare species of
ancient, good-quality forest habitats
in Europe, this particular specimen
was photographed in Austria.

Nosodendridae are a mainly tropical group called diseased-tree beetles, a translation of their scientific name. Adults can often be found on slime fluxes, the flows of fermenting sap on the trunks of living trees, which support a rich community of fungi and insects. It is not clear if they are feeding on the sap itself or perhaps on other inhabitants. There is still much to be discovered about these uncommon beetles.

Jacobsoniidae, named after Russian entomologist Georgiy Jacobson (1871–1926), is an obscure family of tiny beetles, many less than 1 mm long.

were extracted using funnels from leaf litter or other organic detritus, or from bat guano in caves, and many are reported from islands, from Mauritius to Madeira, Christmas Island to the Caribbean. It is probable that they are usually overlooked, and only detected by more specialized and detailed methods of collecting, which are more likely to be used in small, species-poor environments such as islands

POWDER POST BEETLES, WOODWORMS, AND RELATIVES

Bostrichoidea is a small superfamily of Polyphaga, with fewer than 4,000 species, about 400 genera, and four families. All four families have an ability, aided by gut microbes, to develop on dry organic matter; this brings some species into conflict with humans.

Bostrichidae, called augur beetles, includes some 600 species. With a few exceptions (such as the Lesser Grain Borer *Rhyzopertha dominica*, a stored crop pest), larvae bore in wood, mostly when it is dead and dry. Adults of several species can emerge suddenly and unexpectedly from imported ornaments or furniture, inside which larvae have been growing for several years. The largest species, *Dinapate wrightii*, the Giant Palm Borer, is associated with palm trees in California, and reaches lengths of 2 in (5 cm).

Ptinidae includes 2,200 species. Spider beetles (subfamily Ptininae) are named for their round body and long, outspread limbs. Woodworm and death watch beetles (subfamily Anobiinae) include several important pests of furniture and building timber.

Endecatomidae is a small family that contains only four small beetles in the genus *Endecatomus*, found on fungus-infested wood in forests of the northern hemisphere.

Dermestidae, hide beetles or skin beetles, includes around 1,200 species, which mostly feed on dry animal matter. The genus *Dermestes* is common on dry carrion outdoors, and some also infest dried meat or fish products, earning the name "larder beetles." Several genera of dermestids are called "museum beetles," as they are destructive pests of collections of taxidermy, skins, or insects. The bristly larvae, known as "woolly bears," do most of the damage, while the adults feed outdoors on nectar from flowers.

SUPERFAMILY
Bostrichoidea

KNOWN SPECIES
4,000

DISTRIBUTION
Worldwide, except Antarctica. Several species have been spread by human trade

HABITAT
Forests, scrub, human habitations

SIZE
2–50 mm

DIET
Dry organic matter, from dry wood and grain (Bostrichidae, Ptinidae: Anobiinae), to dry animal matter (Dermestidae, Ptinidae: Ptininae)

NOTES
Some Dermestidae, in the genera *Anthrenus*, *Trogoderma*, and *Reesa*, are important pests not only in homes but also in museums, where natural history curators struggle to keep them out of the collections. Museum beetles probably evolved feeding on dry insect

ABOVE | *Apate monachus*
(Bostrichidae) This large African
wood-feeding bostrichid is
spreading into Europe; this picture
represents the first from Hungary.

OPPOSITE | *Anthrenus verbasci*
(Dermestidae) Varied Carpet
Beetle larvae are serious pests
in insect collections in Europe
and have been introduced into
North America.

RIGHT | *Ptinus schlerethi* (Ptinidae)
A typical spider beetle, this species
lives in old forest trees. Similar
species scavenge on dry organic
matter in houses.

exoskeletons discarded by spiders, or in old
wasp nests, but they can reduce a box of
dry insect specimens to dust and labels in
a few months. On the other hand, other
Dermestidae, in the genus *Dermestes*, are
deliberately cultivated by museums in rooms
called "dermestaria," where the flesh-eating
larvae are used to clean the remaining meat
and tissue from bones, leaving perfectly
clean skeletons

SHIP TIMBER BEETLES

The superfamily Lymexyloidea is represented today by only a single family, Lymexylidae. The world fauna includes only a few genera and species, and these are often rare. Evidence shows that they were more common in the geological past, with a number of fossils known from Cretaceous amber, 100 to 120 million years old.

The beetles are wood borers, particularly associated with standing dead wood such as lightning-struck trees. This has earned them the name of ship timber beetles, as the masts of old wooden sailing ships are effectively standing dead trees from a beetle's point of view. They can be pests, because they bore directly into the heartwood, damaging the strength of the timber, a particular problem in the days of sailing ships when infested weakened masts could snap off in a gale.

The female beetle bores a cylindrical tunnel directly into the wood at right angles to the trunk, and there she lays the eggs. Each egg is coated with fungal spores which the female deliberately carries in special pouches called "mycangia." These spores are then "farmed" by the hatching larvae in their own branching tunnels. The fungus, often unique to a species of beetle, thrives in the dark humid network of holes, which are kept open by the beetles to ensure a supply of oxygen. This is an example of symbiosis, where the two organisms, the beetle and the fungus, are mutually dependent. The fungus is carried by the beetles from tree to tree and cultivated in their tunnels, where it provides a source of food for the larvae.

LEFT | *Lymexylon navale* The name refers to the navy and the former association with ships' masts. This is now a rare species in Europe and western Asia.

SUPERFAMILY
Lymexyloidea

KNOWN SPECIES
70

DISTRIBUTION
All continents except Antarctica, but generally rare. One or two species are common in tropical Africa

HABITAT
Forests, from northern conifer forests to tropical rainforests. Larvae usually develop in dry, dead trunks

SIZE
10–50 mm

DIET
Larvae feed on symbiotic fungi inoculated into tunnels in wood by the female. Adult feeding is not known, and possibly many species do not eat as adults

ABOVE | *Elateroides flabellicornis* has striking "flabellate" antennae in the male, for detecting the female pheromones. This European species feeds on conifers.

RIGHT | *Atractocerus* This tropical genus has tiny elytra and a huge abdomen, hardly looking like a beetle at all. This is an Asian species.

NOTES

The taxonomic placement of Lymexyloidea is very difficult, as they seem to have no close relatives in the modern fauna, and have been placed in a superfamily of their own. They are so highly modified that on first inspection, they might not even be recognized as a beetle, especially those species where the elytra are reduced to tiny flaps. DNA studies seem to suggest that they may be related to the Tenebrionoidea

CHECKERED BEETLES AND ALLIES

The Cleroidea is a medium-sized superfamily within Polyphaga, comprising just over 10,000 known species worldwide, in 13 families. However, as with many beetle superfamilies, the species are unevenly spread across the families. Four families, Phloiophilidae, Metaxinidae, Phycosecidae, and Acanthocnemidae, each has only a single genus with one to four species, while Cleridae and Melyridae, covered separately below, each include several thousand and 90 percent of the species between

them. The third largest family of the Cleroidea is the Trogossitidae, known as "bark-gnawing beetles," with about 600 species in 50 genera. Many of these are predators found in the tunnels of bark beetles and other wood-boring insects. Some are useful biological control agents of forestry pests. One species—"the cadelle," *Tenebroides mauritanicus*—was spread all over the world with stored products, though it probably often eats other stored-product insects rather than the crops themselves.

SUPERFAMILY
Cleroidea

KNOWN SPECIES
10,500

DISTRIBUTION
Worldwide except Antarctica

HABITAT
Most habitats, particularly forests. Some species in stored products

SIZE
2–50 mm

DIET
Many species are predators, while others are flower feeders or feed on fungi

NOTES
The only member of the Phloiophilidae, the European *Phloiophilus edwardsii*, is a small beetle 2–3 mm long, attractively marked in brown and black. It occurs on the fungus *Peniophora quercina* growing on the twigs

Other families (with the number of species given in brackets) are: Chaetosomatidae (12), Thanerocleridae (30), Mauroniscidae (26), Prionoceridae (160), Byturidae (24), and Biphyllidae (200). The last two, Byturidae (fruitworm beetles) and Biphyllidae (false skin beetles), were for many years in the superfamily Cucujoidea, but recent DNA work has revealed that they actually belong in Cleroidea.

Byturids, as their English name suggests, develop as larvae within fruits such as raspberries and blackberries, and can be a minor pest. The adults feed on flowers, often clustering on yellow buttercups. Biphyllidae are fungus beetles that develop under bark or on fungal fruiting bodies.

ABOVE | *Tenebroides mauritanicus* (Trogossitidae) Called "the Cadelle," originally a French name thought to be derived (oddly) from the Latin for a puppy.

RIGHT | *Peltis grossa* (Trogossitidae) This large European beetle lives in and around bracket fungi on trees, especially in northern forests.

OPPOSITE | *Idgia* (Prionoceridae) From Southeast Asia, this beetle is a typical member of the small family Prionoceridae, which are diverse in Asian forests.

and branches of oak, and the adults are most active during the winter months, which is unusual among beetles. Taxonomically isolated, not closely related to other beetles, the famous coleopterist Roy Crowson once described this as "the most interesting beetle species in Britain"

CLERIDAE
CHECKERED BEETLES

The family Cleridae includes more than 3,000 species of predatory beetles. Many are strikingly patterned, and they have earned the name checkered beetles because some are marked in an alternating pattern of different-colored squares, which breaks up their outline and camouflages them against lichen, fungus, or moss-covered bark. Many other clerids mimic ants, or stinging flightless wasps called "velvet ants," increasing this resemblance by moving in a jerky, antlike way.

Most clerids are associated with trees, where they feed on wood-boring beetles, and some are beneficial to forestry because they help to control populations of pest bark beetles (Curculionidae: Scolytinae) or woodworms (Ptinidae: Anobiinae). Some Cleridae have moved into other habitats; the metallic blue species of the genus *Necrobia*, for example, lives under dry carrion, feeding on other insects and occasionally on the carrion itself. These are sometimes called "ham beetles," because they may be attracted, especially in the days before the invention of refrigeration, to dried meats such as hams stored for human consumption—but they are much more common under roadkill or sun-dried carcasses. Another genus, the Eurasian *Trichodes*, is a parasite in the nests of solitary bees, where the clerid larva feeds on the larva of the bee; the brightly colored red and blue

LEFT | *Allochotes sauteri* Like many Taiwanese species, this beetle is named after entomologist Hans Sauter (1871–1943), who intensively studied the island's natural history.

FAMILY
Cleridae

KNOWN SPECIES
3,400

DISTRIBUTION
Worldwide, especially in the tropics

HABITAT
Forests, on dead wood. Some species are found in meadows and open plains

SIZE
2–45 mm

DIET
Predators as adults and larvae, often feeding on the adults and larvae of other beetles

NOTES
The carrion-associated clerid beetle *Necrobia violacea* was recently identified from fossil fragments found in California's famous Rancho La Brea Tar Pits, and dated to 44,000 years old. This is interesting

adults visit flowers, where they feed on other pollinating insects, especially flower beetles.

Cleridae usually require warm temperatures and unspoiled habitats; many are listed as vulnerable or endangered in the Red Data Books of European countries, where they are often found only in the best ancient woodlands.

ABOVE | *Omadius* From Southeast Asia, the checked pattern camouflages this beetle against the moss and lichen on a tree trunk.

RIGHT | *Thanasimus formicarius* The Ant Beetle is a common predator in European conifer forests. It is very antlike, especially in the way that it moves.

because the same species has been found in Egyptian mummies, and in wooly mammoth remains from England. This piece of evidence revealed an early connection between the faunas of Eurasia and North America

SOFT-WINGED FLOWER BEETLES

The family Melyridae is called soft-winged flower beetles because the elytra, in many beetles one of the hardest parts of the body, are not very sclerotized; in fact, the whole exoskeleton is quite flimsy. They were combined with the soldier beetles (Cantharidae) and some other superficially similar families in a group called Malacoderms (meaning "soft skinned"), but these families were found not to be closely related.

Adult Melyridae are short-lived, and are often brightly colored in red, yellow, and metallic green, warning of their potential toxicity. Several species, especially in the subfamily Malachiinae, have colorful sacs along the sides of the body, which they inflate as a threat display when disturbed, sometimes combined with a smell. Like many beetles that have chemical protection from predation, they feed openly on flowers and vegetation during the day.

While most are harmless or beneficial, a few species can reach pest proportions, a striking example being the Spotted Maize Beetle *Astylus atromaculatus* from southern South America. This yellow and black spotted beetle is a good pollinator, and at moderate densities may be useful to agriculture. However, especially outside of its native range, it can develop huge populations, at which point the hungry beetles will eat flowers and damage

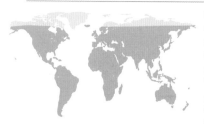

FAMILY
Melyridae

KNOWN SPECIES
6,000

DISTRIBUTION
Worldwide

HABITAT
Common in most habitats, from semideserts to lakesides, even on beaches on the strandline

SIZE
1–10 mm

DIET
Some are predators, others feed on flowers as adults, and may be important pollinators. Larvae are scavengers and predators in soil, leaf litter, or dead wood

NOTES
The brightly colored New Guinean melyrid genus *Choresine* is remarkable for producing batrachotoxins, the same powerful cardio-

RIGHT | *Carphurus* This soft-winged flower beetle has short wing cases, and bright warning colors. This one is feeding on moth eggs.

ears of corn and other crops, while the larvae, scavenging in the soil, will feed on seeds and seedlings. More seriously, they are a veterinary risk, because livestock grazing in pasture heavily infested with adults can be poisoned by ingesting large numbers of the adult beetles.

BELOW | *Malachius bipustulatus* The Malachite Beetle is a colorful species commonly found in summer in flowery meadows throughout Europe.

OPPOSITE | *Astylus variegatus* on flowers in Brazil. These excellent pollinators can occur in huge numbers.

and neurotoxins found in South American poison dart frogs. This poison may deter most predators, but some birds, such as the hooded pitohui from New Guinea's rainforests, also secrete batrachotoxins in their skin and feathers, to the extent that researchers have reported numbness and skin tingling from merely touching live birds. It is thought that these strange "poisonous birds" gain their chemical defense from seeking out and eating the melyrid beetles

CUCUJOID BEETLES

BELOW | *Passandra trigemina*
(Passandridae) A Parasitic
Flat Bark Beetle from southern
Asia, which develops on the larvae
of wood-boring beetles such
as Cerambycidae.

Although only the eighth largest of the superfamilies, with slightly more than 20,000 known species, the Cucujoidea is one of the most variable groups in terms of life history, and is divided into 36 families, more than twice as many as, for example, the similar-sized superfamily Elateroidea (24,000 species in only 17 families). Unsurprisingly, given this diversity, many experts regard Cucujoidea as up to four different superfamilies (Cucujoidea, Nitiduloidea, Erotyloidea, and Coccinelloidea). Some of the cucujoid families, such as the Coccinellidae (ladybugs) are conspicuous, brightly colored, and well known, but many cucujoid families comprise only obscure small, brownish beetles that are difficult to classify, and as a result their taxonomy has been for many years in a state of constant change.

While the majority of Cucujoidea feed as adults and larvae on fungi, either on the fruiting bodies or the spores and mycelia, some have become predators (many Coccinellidae) or phytophagous on living plants (some coccinellids, many nitidulids). Others (many Nitidulidae) are associated with sap or fermenting fruit, and a few, such as the larvae of

SUPERFAMILY
Cucujoidea

KNOWN SPECIES
21,500

DISTRIBUTION
Worldwide except Antarctica

HABITAT
Forests, grasslands, agricultural land, sometimes human habitations

SIZE
1–20 mm

DIET
Many are fungus feeders as adults and larvae, but different families have adapted to a wide range of feeding habits

NOTES
Around three-quarters of all cucujoids belong to just four families: Coccinellidae, Erotylidae, Endomychidae, and Nitidulidae. Most of the other families are small, and several, including Tasmosalpingidae, Priasilphidae, Lamingtoniidae, and Protocucujidae, have an extreme

Passandridae, are among the only beetles that have adopted a parasitic lifestyle, developing on the larvae of other beetles. The Mexican Bean Beetle *Epilachna varivestis* (Coccinellidae), one of the few vegetarian ladybugs, is a minor pest of beans in the USA and Mexico. Some predatory Monotomidae, such as *Rhizophagus grandis*, are used in forestry for biological control of bark beetles (Curculionidae: Scolytinae), such as the conifer pest *Dendroctonus*. Some cucujoids are associated with stored foodstuffs, and members of Silvanidae and Laemophloeidae have been introduced around the world with seeds, grains, and similar products.

ABOVE | *Corynomalus marginatus* (Endomychidae) A cluster in Ecuador; both adults and larvae have conspicuous warning colors to deter predators.

BELOW | *Shoguna* (Monotomidae) A predatory Asian beetle adapted to follow other wood-boring beetles into their burrows and eat them.

southern-hemisphere distribution, being restricted to temperate South America, Australia, and New Zealand. These are probably remnants of the beetle fauna of the ancient southern supercontinent of Gondwana, which broke up during the Jurassic period about 180 million years ago. These small beetles preserve the appearance of some Mesozoic groups, and are not closely related to one another or to other living beetle families

PLEASING FUNGUS BEETLES

BELOW | *Erotylus* Members of this genus are some of the most colorful and conspicuous beetles likely to be seen while walking in the jungles of South America.

Erotylidae, the pleasing fungus beetles, are generally a tropical group, with the adults and larvae feeding together on bracket fungi and mushrooms. Their greatest diversity occurs in wet tropical forests, where high rainfall, humidity, heat, and abundant organic matter result in enormous fungal growth, and many species of erotylid can be found together on the short-lived fruiting bodies of rainforest fungi. Together with their fungal hosts, Erotylidae play an important ecological role as recyclers of complex chemicals such as lignin and cellulose from decaying trees. Erotylidae also secrete chemicals themselves, and their bright warning colors show that they are distasteful to predators. When threatened, they "reflex bleed" by secreting from their joints globules of offensive fluid, which is repellent to vertebrates as well as insect predators such as ants. Like a lot of brightly colored

FAMILY Erotylidae	**DIET** Adults and larvae feed on the fruiting bodies of various fungi. One group, the Languriinae, feed inside the stems of living plants
KNOWN SPECIES 3,500	
DISTRIBUTION Worldwide but especially in the tropics	**NOTES** Some Erotylidae, like the South and Central American genera *Pharaxonotha* and *Ceratophila*, have a completely different way of life from the others, being pollinators of a very ancient group of plants called cycads. They look unlike other erotylids, and probably resemble the ancestral
HABITAT Anywhere that fungi grow, usually in warm, wet forest habitats	
SIZE 1–20 mm	

insects, they make little attempt to hide, and so they can often be found sitting on leaves or flying across forest paths during daylight hours.

One group of erotylids, the subfamily Languriinae, which are also known as lizard beetles, were for many years classified as a different family because their appearance and behavior is quite different from the majority of other erotylids.

The larvae bore in the stems of living plants, and the long, slender adults usually feed on leaves. Like their relatives, the lizard beetles also often display bright warning colors. Both the typical pleasing fungus beetles and the lizard beetles can also be found in temperate environments, especially in fungi and decaying wood, but in smaller numbers than in the tropics.

cucujoids that pollinated cycads during the Jurassic. Many people call the Mesozoic Era the "Age of the Dinosaurs," but to many botanists, it is the "Age of the Cycads"

SAP AND POLLEN BEETLES

Nitidulidae is a medium-sized family of about 4,500 species of generally small beetles, many of which are pollinators and flower feeders, or are associated with fermentation, such as tree sap, or decaying fruit and vegetable matter. A few, such as the genera *Nitidula* and *Omosita*, have moved into decaying animal matter and are found on carrion. Some species feed on the fruiting bodies of fungi, particularly mushrooms.

While many species are ecologically important as pollinators and recyclers, some have become pests. The extremely common Pollen Beetle *Meligethes aeneus* develops as a larva in the flower buds of yellow-flowered plants of the cabbage family, and is a pest of several related crops. The flower-feeding adults are attracted toward anything yellow, and may swarm in huge numbers on yellow cars or yellow clothing. Some farmers try to control

FAMILY
Nitidulidae

KNOWN SPECIES
4,500

DISTRIBUTION
Worldwide

HABITAT
Grasslands, plains, forests

SIZE
1–10 mm

DIET
Larvae of some species develop in the buds of flowers, while the adults are pollinators. Others feed on fermenting vegetable matter, and a few on decaying carrion

NOTES
The tiny (1 mm) spherical beetles of the subfamily Cybocephalinae are often treated as a different family from the Nitidulidae, partly because their behavior is different. They are predators of scale insects (Hemiptera: Sternorrhyncha), which

them by putting out large yellow sticky traps to capture the adults.

Members of the genus *Carpophilus*, feeding on dry fruit, are pests of products such as dates, figs, and raisins, and have been spread throughout the world with trade. Some members of the genus *Stelidota* can attack fresh fruit crops. Some other species are pests of commercial mushroom farms, or of beehives, but the vast majority of species live inconspicuously in the forests and grasslands of the world. Unlike many cucujoid groups, Nitidulidae can be quite species-rich in temperate countries. For example, almost a hundred different species are found in Britain alone.

ABOVE | *Cyllodes* An unidentified species of *Cyllodes* from tropical Asia. One member of this large genus, *C. bifascies*, is a pest of oyster mushrooms.

OPPOSITE | *Aethina tumida* The Small Hive Beetle is an African species that is a serious emerging pest of beehives throughout the world.

RIGHT | *Glischrochilus japonicus* Showing its impressive mandibles; up to 1 cm long, this northeast Asian species is one of the largest Nitidulidae.

are pests of many ornamental and crop plants. Some species, such as the originally Japanese *Cybocephalus nipponicus*, have been spread throughout the world as biological control agents of scale insects on olives and other crops

LADYBUGS (OR LADYBIRDS)

Ladybugs (known as ladybirds in the UK) are among the best-known and best-loved of insects, even if people don't always know that they are, in fact, beetles! These instantly recognizable, hemispherical, and brightly colored little insects move around openly, seemingly unafraid and inoffensive, even in parks and gardens, and often fly onto people. They have become the subject of a lot of folklore, and inspired children's poems, rhymes, and sayings. In many cultures ladybugs are thought to bring good luck, and their name in several languages connects them with divinity. "Ladybird" comes from "Our Lady," the Virgin Mary, and this connection is even more explicit in, for example, the German *marienkafer* (Mary's beetle). Early depictions of the Virgin Mary showed her in a red cloak, and the seven spots of the common red species (*Coccinella septempunctata*) apparently represented her seven joys and seven sorrows. Other translated names include God's cow, Moses's

FAMILY
Coccinellidae

KNOWN SPECIES
6,000

DISTRIBUTION
Worldwide

HABITAT
Gardens and parks, forests, plains, agricultural land

SIZE
0.8–18 mm

DIET
Many species eat aphids and other small insects. Others feed on mildews and similar fungi growing on plants, and a few feed on the leaves of the plants themselves

NOTES
The variable-colored Harlequin Ladybug *Harmonia axyridis* from East Asia is an effective predator of aphids and other small plant pests, and has been introduced to Europe and North America for biological control. In both continents, it has become

cow, and St. Anthony's cow, connecting them with a harmless bovine as well as with a sacred figure.

In reality, the familiar red and black ladybugs are neither harmless nor divine, but adults and larvae are voracious predators of aphids and other small insects. This also makes them popular and beneficial in the garden, where aphids can be a major pest. Most of the other coccinellid species keep a lower profile, and about half of them are called "inconspicuous ladybugs," because they lack the spots and bright colors. While most ladybugs are predators, the subfamily Epilachninae includes leaf-feeding species, and a few of these are even pests of some crops, such as melons and cucumbers.

ABOVE | *Heteroneda reticulata* Not all ladybugs are red with black spots. The Asian predatory Netty Beetle is yellow with a pattern of black lines.

OPPOSITE | *Cryptolaemus montrouzieri* The Australian Mealybug Destroyer, and its wax-covered larva (right), are voracious consumers of scale insect pests, and have been imported globally for biocontrol.

RIGHT | *Epilachna extrema* From South America, this is one of the minority of ladybugs that are herbivorous (leaf feeders) as adults and larvae.

extremely common within the last 20 years, and spread well beyond the glasshouses and gardens where it was introduced. There is a concern that it may outcompete native ladybugs, and it can also enter houses in large numbers to overwinter, which does not endear it to some householders. This species may not be a very good ambassador for a formerly entirely popular group of insects

TENEBRIONOID BEETLES

Tenebrionoidea is the fourth largest of the beetle superfamilies, exceeded only by the "big three" of Staphylinoidea, Chrysomeloidea, and Curculionoidea. It includes over 33,000 species in more than 3,000 genera and 28 families. Tenebrionoids are not well represented in the fossil record, but undoubted tenebrionoid fossils are known from at least the Middle Jurassic, about 170 million years ago. Most of the diversity of the superfamily consists of the darkling beetles (family Tenebrionidae) with over 20,000 species, but many of the smaller families are ecologically and behaviorally interesting.

A large number of them are associated with forests, with larvae developing in dead wood, tree fungi, or as predators under the bark of trees. In fact, 24 of the 28 families have representatives partly or entirely associated with this habitat. Many wood-feeding tenebrionoids play an important role in forest health, and several are of conservation importance in the much-decreased forests of Europe and North America. In many forest species, the larval stage is much longer than the adult life, so adults are rarely seen.

The smallest Tenebrionoidea are the tiny Ciidae (usually less than 2 mm long), which bore in hard, woody bracket fungi, sometimes in huge numbers. The largest are the Trictenotomidae, a small family of sometimes metallic beetles from tropical and subtropical Asia. Reaching 3½ in (9 cm), with long antennae and large mandibles, they are easily mistaken for Cerambycidae or even stag beetles. Like almost all adult tenebrionoids, they can be recognized by the structure of their tarsi (feet), which have one fewer segment on the back legs.

LEFT | *Glipa malaccana* (Mordellidae) Tumbling Flower Beetles, this one from Asia, leap, then drop to the ground if disturbed.

SUPERFAMILY
Tenebrionoidea

KNOWN SPECIES
33,727

DISTRIBUTION
Worldwide except Antarctica

HABITAT
Most habitats, from dry deserts to the edges of glaciers. Many species are associated with forests. Some have moved into stored products and even human habitations

SIZE
1–80 mm

DIET
The diet varies profoundly, but very few Tenebrionoidea feed on living plants. Many are predatory as larvae, but almost none as adults. Adult Tenebrionoidea often feed on pollen and nectar, fungi, and dead plant and animal matter, or in some cases do not eat at all

NOTES
Some of the small families of
Tenebrionoidea, such as Chalcodryidae,
Promecheilidae, and Ulodidae, totaling
fewer than 100 species together, are found
in New Zealand, southern Australia, and
the southern tip of South America (in Chile
and Argentina). This distribution is on the
fragments of the old supercontinent of
Gondwana, which broke up during the
Mesozoic, and shows they belong to
a very ancient group

IRONCLAD AND CYLINDRICAL BARK BEETLES

opheridae is a modest-sized family with fewer than 2,000 named species, although their biology is very variable. In older literature, the name "Zopheridae" referred only to the large ironclad beetles, with the majority of smaller species being placed as "Colydiidae" or "Monommatidae," but the two latter families have now been absorbed into Zopheridae. Most species are fungus or detritus feeders, living under the bark of trees. The true ironclad beetles resemble the others but are much larger, almost 2 in (5 cm) long,

and as the name suggests, are hard-bodied and robust beetles with their elytra fused together, making them very strong but unable to fly. True ironclad beetles are found in most continents, but are most abundant and species-rich in North America, with ten species found in the USA and many more in Mexico.

There is a curious tradition in Mexico of capturing the larger species of the genus *Zopherus* and decorating them with costume jewelry and pieces of brightly colored cloth, attaching them

RIGHT | *Pristoderus chloreus*
This Australian beetle lives among lichens on tree bark. It secretes a coating that resembles lichen, and algae even grow on the beetle.

FAMILY
Zopheridae

KNOWN SPECIES
1,700

DISTRIBUTION
Worldwide except Antarctica. The large ironclad beetles are typical of North America, but some species of Zopheridae can be found in suitable habitats almost everywhere, although they are rarely common

HABITAT
Generally found in forests, associated with living and dead wood

SIZE
2–50 mm

DIET
Most species feed on fungi, detritus, and the larvae of other insects on or under the bark of trees. Some, like the genus *Colydium*, are predators of wood-boring beetles

to a chain, and wearing them as a living brooch called a "makech." These decorated beetles will live quite a long time, even without food and water, and they are often sold as tourist souvenirs, which can result in the unfortunate beetles being confiscated at the US border.

The majority of species of Zopheridae in the broad sense, excluding the true ironclad beetles, are able to fly, and live as fungivores, predators, or scavengers, usually associated with old trees and good-quality forest habitats.

ABOVE | *Zopherus nodulosus haldemani* Found in Texas and northern Mexico, this large, flightless species is a classic true ironclad beetle.

BELOW | *Colydium lineola* A predatory Cylindrical Bark Beetle from North America, well adapted to pursue wood-boring beetles in their tunnels.

NOTES
A large, flightless species from California, the Diabolical Ironclad Beetle *Nosoderma diabolicum* has such a hard exoskeleton that it can survive being driven over by a car. Researchers are studying this beetle in the hopes of designing hard, strong, yet lightweight materials

DESERT BEETLES

BELOW | *Adesmia cancellata*
A typical beetle of dry habitats, from the eastern Mediterranean to the Arabian Peninsula.

Tenebrionidae are often called darkling beetles due to their somber colors and association with dark places, but one large subfamily, Pimeliinae, is associated with the brightest and most relentlessly sunlit environments on earth— the world's deserts and semideserts. Usually called desert beetles, almost all Pimeliinae are xerophilous, meaning they are adapted to dry habitats; in some cases so dry that few other creatures can survive there at all. These adaptations to conserve liquid generally involve fusing of the elytra (reducing evaporation), and the whole subfamily has lost the ability to produce defensive fluids. Instead, adult Pimeliinae are defended by a strong armored exoskeleton. They are built to last and can be very long-lived, up to several years as adults, which is necessary because opportunities to breed and even eat can be limited in the desert environment. They generally eke out a living by scavenging on scraps of plant and animal matter blown along the desert sands, and spend the hotter part of the day buried in the sand, under pieces of debris, or in and under bushes and desert plants.

SUBFAMILY
Pimeliinae

KNOWN SPECIES
8,000

DISTRIBUTION
Worldwide, in warmer parts of most continents (except Antarctica), especially southern Africa

HABITAT
Warm or hot, dry places, especially deserts and semideserts, as well as scrub, savanna, and chaparral

SIZE
3–80 mm

DIET
Detritus. Plant and occasionally animal debris

NOTES
Some toktokkies have become "fog harvesters." Standing on desert sand dunes,

LEFT | *Dichtha cubica* The White-Legged Toktokkie from southern African deserts earns its name from the rhythmic tapping sound it makes with its abdomen.

BELOW | *Lepidochora eberlanzi* An African species strongly adapted for "swimming" through wind-blown sand in the Namib desert.

Some genera, such as *Psammodes*, raise their abdomen and then rhythmically tap it on the ground as a means of communication, making a loud ticking noise. This has earned them the local name in southern Africa of "toktokkies," and a new genus was even named *Toktokkus* in recognition of this behavior. Toktokkies, being flightless, can be very big: the southern African *Stridulomus sulcicollis* can be 3 in (8 cm) long.

with their abdomen raised and their head down, they allow tiny droplets of atmospheric water vapor from the morning mist, which even occurs in deserts, to condense on their elytra and flow down special grooves toward their mouth. This behavior, regularly repeated, allows them to collect enough moisture to survive in the driest deserts, where few insects can live

TRUE DARKLING BEETLES

The Tenebrioninae is the second largest of the subfamilies of Tenebrionidae, and is distributed worldwide. It has been divided into 29 tribes. The genus *Tenebrio* (tribe Tenebrionini) includes the familiar orange-brown-colored mealworm beetles, the larvae of which are bred in huge numbers as food for pets and wild birds, and even made into flour for human consumption.

They are called mealworms because they breed in oatmeal and flour, and were a stored-product pest. This showed that they were preadapted to dry indoor conditions and could be easily reared on cheap, available foods, so they were ideal for domestication.

Some species are associated with human dwellings. For example, the European genus *Blaps* (tribe Blaptini), the Cellar or Churchyard Beetle, was formerly common around stables and granaries but declined with increasing chemical use and the decline of horse-drawn transport. However, the majority of Tenebrioninae are

BELOW | *Ammodonus fossor* Found in sandy forest clearings in continental USA. It may even benefit from occasional fires.

SUBFAMILY
Tenebrioninae

KNOWN SPECIES
7,000

DISTRIBUTION
Worldwide, but particularly the Old World tropics

HABITAT
Forests, also human-altered habitats

SIZE
2–40 mm

DIET
Larvae often in dead wood. Adults are general detritivores and scavengers

NOTES
Many Tenebrioninae have well-developed horns on the head, which are present only in the males, and, as in some other beetle groups, these vary in size and seem to have some role in mate selection. Horns are relatively uncommon in Tenebrionoidea, but occur in several tribes of Tenebrioninae

forest insects. Some, such as the tribe Bolitophagini, are primarily fungus beetles, inhabiting long-lived bracket fungi growing on trees, but the vast majority of species are associated as larvae with dead wood. The tribe Amarygmini, including the genus *Amarygmus*, one of the largest genera in the beetles, comprises over 1,800 described species of mainly brightly colored, metallic, semispherical insects, distributed in Old World tropical forests, especially in Asia and Australasia.

Tenebrioninae adults have defensive scent glands that produce a distinctive and persistent odor when the beetle is threatened. This secretion includes methyl or ethyl phenol, which has a strong taste and can be painful if it comes into contact with the eyes.

ABOVE | *Byrsax* A male of the horned fungus beetle from Malaysia, a member of the tribe Bolitophagini.

RIGHT | *Blaps mortisaga* Once commoner and even an object of superstition, nocturnal Cellar Beetles are now rarely seen around stables and old buildings.

METALLIC DARKLING BEETLES

The family Tenebrionidae are called darkling beetles, and their scientific name is derived from the Latin *tenebra*, meaning darkness or gloom. Both names refer to their nocturnal habits and highlight the somber colors of most species, which are usually brown or black. However, the mainly tropical subfamily Stenochiinae are a conspicuous exception to this rule, not only being colorful, but in some cases brightly iridescent in the colors of the rainbow. Like most metallic beetles, and unlike the majority of tenebrionids, these species are active during daylight, and the coloration, as elsewhere in the beetles, seems to be a thermoregulation adaptation, since the metallic reflective colors absorb heat at low light levels, such as at dawn and dusk, but reflect off excess heat during the hottest parts of the day. The reflection may also help to break up the beetles' outline by reflecting light, shadow, and surrounding objects in the environment, hiding the beetle from potential predators. Like most other tenebrionid beetles, stenochiines also secrete strong-smelling quinones as a defense.

The largest genus in the subfamily Stenochiinae is *Strongylium*, a group of more than 1,400 brightly metallic tropical species. *Strongylium* are elongate, parallel sided, and can be found on the leaves and branches of tropical trees. The larvae bore in the dead wood of branches and trunks, and the adults browse on fungi and lichen, and can quite frequently be collected using a beating tray.

LEFT | *Tetraphyllus* A hemispherical stenochiine on the bark of a forest tree in Singapore. These beetles graze algae on tree bark.

SUBFAMILY
Tenebrionidae: Stenochiinae

KNOWN SPECIES
2,500+

DISTRIBUTION
Worldwide except Antarctica, but almost all species are found in the humid tropics and subtropics

HABITAT
Most species inhabit trees in wet tropical forests

SIZE
5–40 mm

DIET
The larvae of most species feed in dead wood or fungal fruiting bodies. Adults browse algae and fungi on tree bark

NOTES
Some species of *Strongylium*, like the one illustrated bottom right, show behavioral mimicry of tiger beetles, which are fast and difficult to catch, and also have a strong

ABOVE | *Cuphotes erichsoni* This South American beetle mimics the unpalatable fungus beetle *Gibbifer* (Erotylidae) which occurs in the same environment.

RIGHT | *Strongylium* This Singaporean beetle has two projections on its thorax that resemble the eyes of a fast-running predatory tiger beetle (Carabidae: Cicindelinae).

bite—these *Strongylium* can run a lot faster than one would expect from the usually sedate Tenebrionidae. Others mimic Erotylidae, the pleasing fungus beetles, which can often be distasteful or even toxic. Even those stenochiines which are probably not mimics can strongly resemble and be confused with other beetle families, such as Chrysomelidae or Coccinellidae

195

OIL BEETLES AND BLISTER BEETLES

The Meloidae, a widespread group of Tenebrionoidea, have some of the most intricate life histories of any beetles. In the genus *Meloe*, for example, the female lays hundreds of eggs that quickly hatch into small active larvae called "triungulins," which climb up onto flowers, where they wait for a pollinating insect to arrive. If the pollinator is a solitary bee of the correct species (the host species of the *Meloe* beetle in question), they will climb onto it and be carried back to its nest, where they will kill the egg or larva of the bee and feed on the supplies that have been stored for the bee larva's development. During this process they molt into a large helpless larva that is entirely dependent on the food stored by the bee. If the triungulins attach to the wrong species of bee, or to a wasp or beetle, they will probably starve to death, so the *Meloe* depends on large bee populations of the correct species in order to survive. As a result, they are never common, but their triungulins may

RIGHT | *Horia debyi*
This striking species from the Asian tropics, in this case Borneo, develops in the nests of Carpenter Bees (*Xylocopa* spp.)

FAMILY
Meloidae

KNOWN SPECIES
3,000

DISTRIBUTION
Worldwide except Antarctica, especially in warm, dry places

HABITAT
Most habitats, but most common in savanna with abundant potential hosts

SIZE
10–60 mm

DIET
Larvae may be "kleptoparasites," killing a host larva and feeding on the supplies provisioned for it. Others feed on grasshopper egg cases. Adults feed on nectar, or flowers or small leaves

be carried long distances on the flying bees, enabling, for example, these flightless beetles to colonize islands.

Several Meloidae secrete extremely toxic compounds as defense, for example the Spanish Fly—actually a large, metallic green meloid *Lytta vesicatoria*—secretes the toxic terpenoid cantharidin, which is poisonous to vertebrates. In the eighteenth century it became briefly popular, partly due to the recommendations of the notorious French nobleman Marquis de Sade (1740–1814) as a supposed aphrodisiac, but it is dangerous and probably ineffective. In fact, some cantharidin-secreting meloids, for example the genus *Epicauta*, can be so harmful if ingested that they are associated with livestock mortality caused by grazing in areas where the meloids are abundant.

LEFT | *Mylabris* The larvae of this African blister beetle develops in the egg pods of grasshoppers buried in the ground.

RIGHT | *Meloe proscarabaeus* A female of a European Oil Beetle, which will produce hundreds of triungulin larvae, only a few of which will ever survive to become adults.

NOTES
In some parts of Africa, meloids of the genus *Mylabris* can be beneficial to farmers as their larvae consume the egg pods of locusts, keeping these serious pest grasshoppers under control. However, the adults of the same meloids can be harmful to the same farmers by defoliating or eating the flowers from crops such as millet

ANTLIKE FLOWER BEETLES

The Anthicidae is a medium-sized family of small, fast-moving, generally ground-living beetles in the superfamily Tenebrionoidea. They are known as antlike flower beetles, though most species are not especially antlike, nor are they particularly associated with flowers. The majority seem to be generalist scavengers, small predators,

BELOW | *Notoxus* This widespread genus of beetles has a distinctive thoracic horn, and is strongly attracted by cantharidin secreted by other beetles.

FAMILY
Anthicidae

KNOWN SPECIES
3,000

DISTRIBUTION
Worldwide except Antarctica

HABITAT
Found in natural and human-altered habitats, in leaf litter, vegetable debris, compost, and the rainforest canopy

SIZE
2–17 mm

DIET
Scavengers, detritivores, and predators of small invertebrates

NOTES
Although they do not secrete it themselves, some Anthicidae are strongly attracted by the toxin cantharidin, which is secreted as a defense mechanism by other tenebrionoid beetles. Adult males of *Notoxus monoceros*

and detritus feeders. Some species can be found in very large numbers in habitats such as riverbank vegetation, reed litter, refuse under haystacks, or the housing of domestic animals such as chickens and pigs. Several, such as *Stricticomus tobias*, *Omonadus formicarius*, and *Omonadus floralis*, are "tramp species" that have been shipped around much of the world with human trade. Other species such as *Anthicus watarasensis* and *Cyclodinus salinus* are rare or endangered in parts of their range, being associated with scattered and declining seaside habitats such as shingle banks and salt marshes. The subfamily Notoxinae are characterized by a hornlike projection of the thorax that covers the head from above.

In the tropics, Anthicidae can be larger, but still rarely exceed 1 cm. Some tropical species live in the rainforest canopy and are brightly colored. Some of the largest Anthicidae relatives are in the genus *Ischalia*, which is now usually placed in a family of its own, the Ischaliidae, or at least as the subfamily Ischaliinae within anthicids. *Ischalia* are flattened beetles distributed, rather strangely, in Asia and North America. Many are metallic blue, and several, especially in east Asia, have lost the ability to fly. A new flightless species, *Ischalia akaishi*, was recently described from high in the Akaishi Mountains of Honshu, Japan.

RIGHT | *Anthelephila cyanea* An abundant species in the Asian tropics, on a forest leaf in Singapore.

can be attracted in large numbers to dead Meloidae adults or artificial cantharidin sources, and they have also been observed harassing adults of Oedemeridae to make them produce droplets of defensive cantharidin, which the anthicids then quickly consume. It is thought the anthicids use the chemical in their own mating, passing it to the females as a "nuptial gift" and defense for the eggs

LEAF AND LONGHORN BEETLES

BELOW | *Agathomerus* (Megalopodidae) From the Atlantic Forest of Brazil, many megalopodids are stem borers as larvae.

The superfamily Chrysomeloidea includes the following seven families, with estimated number of known species: Oxypeltidae (3), Vesperidae (75), Disteniidae (336), Cerambycidae (30,079), Megalopodidae (350), Orsodacnidae (40), and Chrysomelidae (32,500). It is the largest superfamily in the Polyphaga, with more than 60,000 species, as it includes the fourth and fifth largest families of Coleoptera (covered separately below). Of the five smaller families, Oxypeltidae, Disteniidae, and Vesperidae are superficially similar to Cerambycidae, while Orsodacnidae and Megalopodidae have been classified in the past as subfamilies of Chrysomelidae. An accurate classification of these seven families, however, remains elusive, so most specialists prefer to regard them as separate families for the time being.

Vesperidae includes some of the strangest Chrysomeloidea. The European genus *Vesperus* has a flightless female that is physogastric, with a swollen abdomen filled with eggs, and the larvae develop in the soil and can even be a minor pest. The Brazilian *Hypocephalus armatus* is even stranger: also flightless, it does not even look like a beetle, but is superficially similar to a mole cricket of the order Orthoptera. They burrow in the ground and their life cycle, as well as their taxonomic placement,

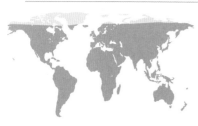

SUPERFAMILY
Chrysomeloidea

KNOWN SPECIES
63,383

DISTRIBUTION
Worldwide except Antarctica

HABITAT
Chrysomeloids are found in most habitats where plants occur, as there are hardly any plants that don't have a chrysomeloid feeding on them

SIZE
1–165 mm

DIET
Chrysomelids and their relatives eat living plants. Cerambycids and related families generally feed as larvae in wood, some in dead wood, others living

NOTES
Chrysomeloidea includes the biggest of all beetles: the Titan Longhorn *Titanus giganteus* (Cerambycidae, subfamily Prioninae) from

remains poorly understood. Oxypeltidae consists of just three species from temperate South America, while Disteniidae is found sparingly throughout the tropics.

The Megalopodidae are uncommon, but species occur throughout the world, especially in warm places. Some species of the genus *Zeugophora* are found in the northern temperate zone, as are several species of *Orsodacne* (Orsodacnidae).

northern South America has been measured up to 6½ in (16.5 cm), larger than a pet hamster! The larva of this species, probably living among the roots of giant rainforest trees, has never been reported. As beetle larvae are usually bigger than the adults they become, this might mean that the largest beetle in the world has not yet been scientifically documented

FLOWER LONGHORNS

The Lepturinae, called flower or blossom longhorns, are a comparatively small subfamily of Cerambycidae, with only about 1,500 species. They are rare in the tropics, but in temperate regions of the northern hemisphere they are abundant and species-rich, so despite the globally small size of the superfamily, it includes some of the commonest and most frequently seen longhorns in Europe, North America, and northern Asia.

The adults of most species are diurnal, fly readily, and feed openly on flowers during the day. They prefer white flowers, either umbellifers (family Apiaceae) or the blossom of trees, and they are important pollinators. They have a characteristic shape, with broad shoulders, triangular pointed elytra, a bell-shaped pronotum, and relatively short antennae for longhorns. Many species have a yellow or orange and black pattern, superficially resembling stinging Hymenoptera (bees and wasps), and they also look very wasp-like in flight. These species are "Batesian Mimics," named after the nineteenth-century English naturalist Henry Walter Bates, who, working with Amazonian butterflies, described how harmless species gain protection by resembling those that are either toxic or harmful.

SUBFAMILY
Lepturinae

KNOWN SPECIES
1,500

DISTRIBUTION
Worldwide except Australia and Antarctica. Rare in the tropics, most species are found in the temperate northern hemisphere

HABITAT
Forests and forest edges

SIZE
5–50 mm

DIET
Larvae usually in dead wood, a few on soil fungi or living plants. Adults feed on pollen and nectar

NOTES
Although completely absent from Australia, and scarce in the tropics, this group is extremely abundant and species-rich in the temperate northern hemisphere, with many

OPPOSITE | *Desmocerus palliatus*
This North American beetle feeds
on living elder wood, and is
probably a mimic of distasteful
net-winged beetles (Lycidae).

ABOVE | *Rhagium bifasciatum*
A common European spring
species; larvae feed under
the bark of dead deciduous
or coniferous wood.

Larvae of most Lepturinae feed on well-decayed
wood, and hence few species are pests of timber.
Probably the larvae get most of their nutrition from
fungi in the wood, and unusually for a longhorn,
the European *Pseudovadonia livida* has abandoned
wood feeding altogether, developing instead on
fungal mycelia in the soil; it feeds specifically on the
underground parts of the fairy ring mushroom,
a common fungus in grasslands and meadows.

species known from Europe, Russia, China,
Japan, Canada, and the USA, extending to
northern latitudes. Lepturinae is only about
4 percent of the world longhorn fauna, but
in Britain, for example, it comprises more
than 30 percent of the fauna. New species
are still being discovered in eastern Asia,
especially in the mountains of China

FLAT-FACED LONGHORN BEETLES

With more than 20,000 species named to date, the Lamiinae, known as flat-faced longhorn beetles because of the flattened shape of the head between the antennae and the mandibles, comprise the largest subfamily in the longhorn beetle family Cerambycidae, and one of the major radiations of species and genera in the Coleoptera. Most species develop as larvae in wood, either living or dead, and in some cases development takes many years, particularly when the wood is dry and low in nutrients. Some genera (for example, *Agapanthia*) develop in the stems of living herbaceous plants such as lilies and thistles, while others (such as *Dorcadion*) live in roots in the soil. The lamiines *Batocera wallacei* from Asia, *Acrocinus longimanus* from South America, and *Petrognatha gigas* from Africa are among the largest of all beetles.

Most lamiines attack freshly fallen or even living wood, and generally they are associated with only a few families or genera of trees. As timber feeders, some species can be pests of forestry. The sawyer beetles of the genus *Monochamus* attack a range of softwoods across the northern hemisphere, and the Asian Longhorn *Anoplophora glabripennis*, originally

RIGHT | *Phytoecia cylindrica* from Europe bores in the stems of Queen Anne's Lace plants. The adults can be seen in early spring.

SUBFAMILY Lamiinae	**SIZE** 3–150 mm
KNOWN SPECIES 20,000+	**DIET** Larvae usually in wood, some taxa in herbaceous stems or roots. Adults may feed on bark, leaves, sap, or in many cases do not feed at all
DISTRIBUTION Worldwide except Antarctica, concentrated in the tropics	
HABITAT Tropical and temperate forests, some genera are associated with grasslands and plains, feeding in soil or stems of herbaceous plants	**NOTES** The Amazonian lamiine genus *Onychocerus*, notably *O. albitarsis*, has the last segment of its antennae adapted into a sting resembling

from China, is a serious invasive pest of ornamental trees in many North American cities. The majority of lamiines, however, inhabit the tropical forests of the world, and are particularly species-rich in the Asian wet tropics and in the Amazonian rainforests of South America. Up to 100 new species are discovered and named by scientists every year.

TOP | *Acrocinus longimanus* The striking pattern of the giant (up to 75 mm) Harlequin Beetle camouflages with lichen and moss-encrusted bark of South and Central American trees

that of a scorpion. This is linked to a venom gland, and the effect is said to be similar to a bee sting. It is the only known example of a stinging beetle

ABOVE | *Macrochenus guerinii* First named from specimens collected in what is now Bangladesh, this striking diurnal lamiine is a familiar sight in forest margins across much of India, China, and Indochina.

CAPRICORN BEETLES

The subfamily Cerambycinae includes a third of the longhorn beetle family Cerambycidae, and with over 11,000 species, is exceeded in species richness only by the subfamily Lamiinae. Like most other longhorns, larvae of Cerambycinae feed inside plant material, mostly woody plants, while the adults feed on tree sap or on nectar from flowers. Some cerambycines prefer living wood, and may be harmful or even fatal to the trees in which they develop. One example is the Australian genus *Phoracantha*, which attacks eucalyptus trees; two species, *Phoracantha semipunctata* and *P. recurva*,

RIGHT | *Xoanodera striata* Only discovered in 1970, in Laos, this beetle is usually found on cracked, mossy old tree bark, where it blends in perfectly.

SUBFAMILY
Cerambycinae

KNOWN SPECIES
11,200

DISTRIBUTION
Worldwide except Antarctica. Most common and species-rich in tropical forests

HABITAT
Forests, woodlands, or wherever there is timber for the larvae to develop in. A few species live in herbaceous plants in grasslands

SIZE
6–150 mm

DIET
Larvae develop inside either dead or living plant material. Adults feed on nectar and pollen or tree sap, and some adults do not feed at all

NOTES
Dead adults identified as *Cerambyx cerdo*, the Great Oak Capricorn, were found inside a piece of buried ancient bog oak in eastern

LEFT | *Rosalia alpina*
The Alpine Longhorn is one of the most beautiful of European beetles. Historically a rarity, it has become more common in recent decades.

have been spread all over the tropics and subtropics with eucalyptus, and can now be found in Africa, South America, southern Europe, and parts of Asia. As they only attack an introduced, non-native, fast-growing tree, it is a matter of perspective whether they are considered a pest or a biological control agent.

The subfamily includes large and brightly colored species, mostly found in tropical forests, but a few extend into northern latitudes. For example, the blue and gray *Rosalia alpina* of Central Europe,

a rare and protected species of warm beech forests, is often considered one of the most beautiful of all beetles. The tribe Clytini includes many wasp beetles that are mimics of stinging Hymenoptera, not only resembling them physically, but mimicking their jerky movements and the movement of their antennae. Bright metallic green or blue members of the tribe Callichromatini are sometimes called musk beetles, as the adults can produce a strong musky smell when handled.

RIGHT | *Necydalis ulmi* A mating pair of this bizarre, wasp-like European beetle. Species in this genus and related genera are sometimes placed in their own subfamily, Necydalinae.

England, and carbon dated as almost 4,000 years old. This large beetle is no longer found in Britain at all, so these ancient dead beetles can teach us about the climate and environment in prehistoric times

TORTOISE BEETLES AND RELATIVES

Cassidinae was recently expanded to include the Hispinae, which were formerly treated as a subfamily in their own right. Cassidinae is a large, cosmopolitan group of plant-feeding beetles within the leaf beetle family Chrysomelidae.

The true tortoise beetles, the tribe Cassidini, are named for their tortoise-like shape. They can defend themselves by clamping down on a leaf surface using specialized pads on their feet, which makes it difficult for a predator to dislodge them or to access their vulnerable underside. Many species show advanced parental care, where the mother protects the eggs in this way from predators and parasitoids such as ants and parasitic wasps. As the larvae grow, the mother continues to protect them until they are ready to pupate. Some species are green or brown to camouflage with their host plants, or to resemble spots of mildew on the leaves. Others are brightly colored to indicate that they are toxic or distasteful, having accumulated chemicals from host plants. Some tropical species are glittering metallic gold or silver, imitating droplets of water on leaves, an effective defense in a wet tropical rainforest.

The Hispini have different defenses: some are armed with formidable spines that would make most birds or amphibians think twice about eating them. These are a primarily tropical group, but several species occur in Europe and North America, many of which are leaf miners in the leaves of various grasses, palms, and other plants.

SUBFAMILY
Cassidinae

KNOWN SPECIES
6,000–7,000

DISTRIBUTION
Worldwide except Antarctica

HABITAT
Tropical rainforests to parks, roadsides, and gardens

SIZE
2–35 mm

DIET
Adults and larvae feed on leaves, many species are host-specific on only a single genus or species of plants

NOTES
Larvae of true tortoise beetles (tribe Cassidini) have moveable spines at the tip of their abdomen that they hold over their back like an umbrella, and on which they collect their own droppings as a means of camouflage and protection. This mobile defensive structure is called a "fecal shield"

ABOVE | *Stolas* Among the largest of the true tortoise beetles, members of this genus can exceed 2 cm in diameter. This species is from the Andes of South America.

OPPOSITE | *Aspidimorpha miliaris* One of the most common of the Asian tortoise beetles, this impressive species can be found in city parks and gardens in tropical Asia.

RIGHT | *Dactylispa* These South Asian spiny leaf beetles are protected from predators, but mating is a difficult procedure. The larvae are leaf miners.

POT BEETLES

The Cryptocephalinae are a medium-sized group of leaf beetles found in a wide range of environments throughout the world, but are most successful in dry, arid habitats such as maquis, garrigue, chaparral, cerrado, and semidesert.

The scientific name means "hidden head," because the head of the adult is retracted beneath the pronotum and not very visible from above. They earn the common name pot beetles because the larvae build themselves a case out of their own

SUBFAMILY
Cryptocephalinae

KNOWN SPECIES
5,000

DISTRIBUTION
Worldwide except Antarctica

HABITAT
Usually dry environments, such as plains, scrub, and semidesert. Some species occur in tropical forests

SIZE
1–12 mm

DIET
Larvae feed on general detritus in the leaf litter or upper soil layer, sometimes in the nests of ants. Adults are usually host-specific on one or several genera of plants

NOTES
Since most Cryptocephalinae are associated with warm, dry environments, species that occur farther north can be quite rare and

OPPOSITE | *Chlamisus* This South American species can retract its head and limbs and resemble the dropping of a large caterpillar, to deceive predators.

ABOVE LEFT | *Cryptocephalus sericeus* A metallic green European species that pollinates a range of plants with yellow flowers.

ABOVE RIGHT | *Cryptocephalus* Larva of a species from Singapore, showing its protective case, or pot, which is made from its own feces.

feces, which resembles a clay pot. They carry this pot like the shell of a hermit crab or a snail, enlarging it as they grow, and it protects them from desiccation, physical damage, and predators, as they can retreat within it and close the entrance with a flat plate on their head. In many species, the larvae allow their pots to be collected by foraging

localized. More than three-quarters of the 24 species occurring in Britain, for example, have been given conservation status and are restricted to just a few known sites. The dry parts of the Mediterranean, Central Asia, and southern North America, on the other hand, support large numbers of individuals and a high diversity of species

ants, then live associated with the ants, scavenging in and around their nests.

Although the larvae are often generalist feeders on dead vegetation, the adults are usually quite host-specific on living plants, often tough xerophytic and aromatic dry-habitat species, feeding on the leaves and accumulating plant compounds for their own defense. Adults of some species also feed on and pollinate flowers, and are diurnally active, so bright metallic colors are common.

When threatened, adult Cryptocephalinae immediately retract their head, antennae, and legs, and drop to the ground. Adults of the tribe Chlamisini imitate the droppings of large caterpillars, which often share the same host plant, to avoid being eaten by birds.

LEAF AND FLEA BEETLES

The Galerucinae, including the jumping flea beetles (tribe Alticini), includes more than 15,000 named species, making it one of the largest beetle subfamilies, accounting for almost half of the diversity of the enormous leaf beetle family (Chrysomelidae).

Almost all galerucines feed on living plants, the adults eating the leaves and the larvae roots or stems. Many species are pests of agriculture or

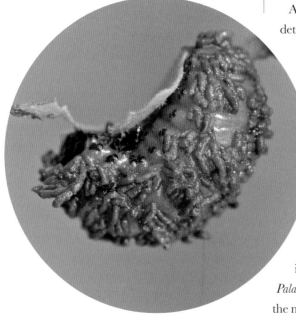

horticulture, while others are important as biocontrol agents of invasive weeds. Some genera, such as the widespread *Longitarsus* and the New World *Diabrotica*, include hundreds of species, and accurate identification can be challenging and involve dissection of adults. This diversity is driven by host-specificity; the majority of galerucines are associated with only a single genus or species of plant, and have evolved very closely with their hosts, so they are able to metabolize toxic anti-herbivory chemicals the plants produce in defense.

A few genera feed on mosses or even general detritus; these include some of the smallest flea beetles at only 1 mm long, and new species are still being discovered in the leaf litter of forest floors worldwide. Another recent discovery, which breaks new ground in leaf beetle behavior and adaptation, is the amazing genus *Myrmeconycha*, named in 2017; the four South and Central American species are myrmecophilous, meaning they live with colonies of ants, a behavior well documented in some beetle groups but not in leaf beetles. The largest Galerucinae is *Palaeosastra gracilicornis*, which reach 20 mm, from the montane tropical rainforests of New Guinea.

SUBFAMILY
Galerucinae

KNOWN SPECIES
15,500

DISTRIBUTION
Worldwide, especially in the tropics

HABITAT
Agricultural fields, plains, temperate and tropical forests

SIZE
1–20 mm

DIET
Adults eat leaves and larvae the stems and roots of living plants, in almost all cases flowering plants

NOTES
The larvae of the African flea beetles of the genus *Diamphidia* collect a toxin from their host plants, called diamphotoxin, and store it in their bodies. Larvae and pupae are traditionally collected by the San Bushmen of the Kalahari Desert, and used to poison arrowheads for hunting, in the same way that

LEFT | *Galeruca tanaceti*
In late summer, females of this European beetle become swollen with eggs. Larvae feed on tansy and bedstraws.

RIGHT | *Diabrotica viridicollis* From Ecuador, this is a typical member of this huge genus. Many North American *Diabrotica* species are crop pests.

OPPOSITE | *Diamphidia vittatipennis* The larva of the African Arrow Poison Beetle stores toxins from its poisonous food plants. It coats itself with its toxic feces for extra protection.

Amazonian people use poison arrow frogs. The chemical is harmless when eaten, but injected into the bloodstream is a powerful and rapid hemolytic poison to mammals

WEEVILS

The superfamily Curculionoidea, the weevils, are characterized by having their head extended between the eyes into a projection called a rostrum, with the mouthparts at its tip. The length and breadth of the rostrum varies, and in some groups (such as bark beetles, Curculionidae: Scolytinae) it has been lost altogether. The rostrum is used as a drill to make a hole into which the egg is laid, enabling the larva to develop deep inside the plant substrate protected from predators, parasites, and dehydration.

Curculionoidea is one of the hyperdiverse groups in the suborder Polyphaga, and is a candidate for the most species-rich beetle superfamily, the other two being Chrysomeloidea and Staphylinoidea, each with more than 61,000 formally named, living species. In the present classification, seven families are recognized in Curculionoidea: the

SUPERFAMILY
Curculionoidea

KNOWN SPECIES
62,000

DISTRIBUTION
Worldwide, including sub-Antarctic islands but not Antarctica itself. Most abundant in the tropics

HABITAT
Most habitats, from deserts to tropical swamps

SIZE
1–75 mm

DIET
Adults and larvae of almost all species feed on plants

NOTES
The family Curculionidae, with more than 50,000 species, is extremely diverse, and several groups now regarded as subfamilies have been treated as separate families in the recent past. One example is the palm weevil

OPPOSITE | *Cercidocerus indicator*
(Curculionidae: Dryophthorinae)
This male of Southeast Asia
shows impressive "windscreen-
wiper" antennae.

ABOVE | *Rhinotia bidentata*
(Belidae) From Australia,
this ancient lineage of
approximately 375 named
species is most abundant in
the southern hemisphere.

Nemonychidae, Anthribidae, Belidae, Caridae,
Attelabidae, Brentidae, and Curculionidae. Of
these, the Curclionidae is by far the largest, with over
50,000 species.

   With only one or two exceptions (for example,
the curculionid genus *Ludovix* feeding on
grasshopper eggs, the anthribid genus *Anthribus*
feeding on scale insects), curculionids are plant
feeders as adults and larvae, and it is thought that
they owe their diversity to a radiation of species in
the Cretaceous period, at the same time that the
flowering plants (angiosperms) achieved global
dominance. However, weevil fossils date back
further than the Cretaceous period, and were
already numerous in the late Jurassic, possibly even
earlier. Weevils remain an ecologically important
group today, and are particularly abundant in
tropical forests.

group Curculionidae: Dryophthorinae,
which has around 1,200 species, including
the largest of all weevils, the Giant Palm
Weevil *Protocerius colossus* at 3 in
(7.5 cm). Dryophthorinae specialize on
monocotyledonous plants, which include
palms, grasses, pineapples, and bananas.
Several have become pests, notably the Red
Palm Weevil *Rhynchophorus ferrugineus*,
the Grain Weevil *Sitophilus granarius*, the
Rice Weevil *Sitophilus oryzae*, and the
Banana Weevil *Cosmopolites sordidus*

ANTHRIBIDAE
FUNGUS WEEVILS

The Anthribidae is one of the most ancient lineages of weevils, with fossils known from the Jurassic of Kazakhstan, more than 150 million years ago, which, in spite of their great age, quite closely resemble species and genera living today. Like many of the early weevil groups, anthribids are characterized by having straight antennae (without the geniculate knee bend that characterizes true Curculionidae) and by not feeding on living plants. Some species have developed very unusual feeding habits: *Araecerus fasciculatus* is a cosmotropical stored-product pest, frequently attacking stored coffee beans and cocoa pods in conditions of high humidity. It is probably digesting fungi, but it does physical damage by boring into the beans and pods. The European

FAMILY
Anthribidae

KNOWN SPECIES
6,000

DISTRIBUTION
Worldwide except Antarctica, but concentrated in the tropics

HABITAT
Forests and woodlands

SIZE
2–45 mm

DIET
Fungi and fungoid wood, a few species feed on stored products and even fewer are predators of scale insects

NOTES
In some Asian tropical anthribids, such as the genera *Mecocerus* and *Xenocerus*, males have antennae many times as long as the body. They are used for mate-guarding, to detect the approach of another male, as in some longhorn beetles (Cerambycidae), which they superficially resemble

genus *Anthribus* is one of the few examples of
a carnivorous weevil, as adults and larvae are
predators and parasitoids of scale insects
(Hemiptera: Coccoidea) and their eggs.

However, the majority of Anthribidae are
associated as adults and larvae with fungi or with
fungoid dead wood, and they can be abundant in
tropical forests throughout the world. A smaller
number of species can be found in temperate
woodlands, but they are almost entirely forest
insects. Most species are patterned in gray, brown,
and white, breaking up their outline and making
them difficult to see. Several species can run and
fly rapidly when disturbed.

ABOVE | *Habrissus ramosus*
This large-eyed and
fast-moving fungus weevil
from tropical Asia was
described as new for science
as recently as 1997.

BELOW | *Euparius* A nocturnal
fungus weevil of the Brazilian
Atlantic Forest. Many
Neotropical fungus beetles
share this speckled pattern.

OPPOSITE | *Platystomos albinus* A fungus weevil
from northern Europe, with a brown and
white pattern that resembles a fleck of fungoid
bark or possibly a bird dropping.

LEAF-ROLLING WEEVILS

Attelabidae, like Anthribidae, Brentidae, Belidae, and Nemonychidae, belong to the group of primitive weevils marked by having straight antennae. They are a relatively small group, but have a long fossil history dating back to the Early Cretaceous. The concept of Attelabidae featured here includes the closely related Rhynchitinae, which are treated by some specialists as a separate family.

Some members of both subfamilies are remarkable for their leaf-rolling habits. The adult female partially cuts a living leaf that is still attached to the tree and folds it into an intricate cigar-shaped leaf roll, in which she lays a single egg.

FAMILY
Attelabidae (including Rhynchitinae)

KNOWN SPECIES
2,500

DISTRIBUTION
Worldwide except Antarctica

HABITAT
Forests, temperate open woodland to tropical rainforest

SIZE
2.5–20 mm

DIET
Plant tissue, larvae on recently dead leaves (in leaf rolls) or living buds and fruits. Adults eat leaves, in some species by scraping off the epidermis

NOTES
Some species of attelabids, such as the European *Lasiorhyncites sericeus* (Rhynchitinae), do not make their own leaf rolls, but their larvae live as "inquilines," or parasites, in the rolls made by other species, in this case the Oak Leaf-roller *Attelabus*

LEFT | *Apoderus*
This is a species
from Malaysia,
but similar-looking
Apoderini occur
worldwide. Most
of them make tight,
cigar-shaped rolls from
the leaves of a single
species of tree.

BELOW | *Byctiscus betulae*
This beautiful European
species is not very
host-specific and
rolls the leaves of
multiple species
of trees and shrubs.
It can be a pest
of grapevines.

The female then cuts off the supply of nutrients to the leaf, so it begins to wither and die, and the larva is able to feed on the dying leaf, while protected from predators, parasites, and desiccation inside the leaf roll. Some attelabines roll numerous leaves on the same tree, making it look like the tree is covered with small fruits. The leaf rolls are distinct, and it is often possible to recognize which species of attelabid has constructed them.

Some members of Rhynchitinae develop in buds, and can be pests by damaging blossom, an example being the North American "rose curculio" *Merhynchites bicolor*, which attacks the buds of cultivated roses. Others, such as *Rhynchites bacchus* and *R. auratus* develop inside stone fruits such as apricots, cherries, and sloe berries, and are regarded as minor pests.

OPPOSITE | *Trachelophorus giraffa*
This large, red and black species shows
marked allometric growth of the head
and thorax, and the elongated neck
(hence the name Giraffe Weevil) is used
in competitions between males.

nitens (Attelabinae). The "thief weevils" of
North America (Attelabinae: Pterocolini)
show similar behavior

PRIMITIVE WEEVILS

Brentidae are sometimes called "primitive weevils" because they have straight antennae (while the other very large family in the superfamily Curculionoidea, the Curculionidae, have antennae that are angled). There are two large subgroups of Brentidae, the subfamilies Brentinae and Apioninae, as well as a number of smaller subfamilies.

The subfamily Brentinae usually have elongated, dark-colored adults, and larvae that develop feeding on fungi inside dead wood. They are almost entirely confined to the tropics and subtropics, and very uncommon in temperate zones; only two species reach southern Europe, three are found in the USA, and one in New Zealand. One of the European species, *Amorphocephalus coronatus*, is extremely unusual among weevils, as it lives and breeds as a scavenger in the nests of ants.

Apioninae, which are small, pear-shaped weevils that feed on living plants and have larvae that develop in seeds or seedheads, are much more widespread, with hundreds of species distributed

FAMILY
Brentidae

KNOWN SPECIES
4,000

DISTRIBUTION
Worldwide except Antarctica. Subfamily Brentinae mainly in the tropics, subfamily Apioninae widespread

HABITAT
Subfamily Brentinae generally in tropical forests, breeding in fallen wood. Apioninae occur in a wide range of habitats depending on their host plant, but they are characteristic grassland weevils

SIZE
1.5–90 mm

DIET
Brentinae larvae feed on fungi growing in their tunnels in dead wood, or in some rare cases, such as *Amorphocephalus*, feed in the nests of the ants with which they live. Apioninae feed on living plants or in a few

throughout the world. As they are host-plant specific, some are minor crop pests or potential vectors of crop diseases, while others have been used with some success for biological control of invasive weeds such as gorse and broom. The Sweet Potato Weevil *Cylas formicarius* is a significant pest of sweet potato throughout the warmer regions of the world.

The brentid subfamily Ithycerinae has only one living species, the large, gray and white New York Weevil, *Ithycerus noveboracensis*. Found only in eastern USA and southern Canada, it seems to have no known close relatives.

cases on plant galls, and the larvae, which develop inside the plant, are restricted to a single species or a few related plant species

NOTES
Males of some species of Brentinae display remarkable size variation and are among the few groups of weevils where males compete with one another. The females are smaller and often lack the exaggerated ornamentation of some males

LONG-NOSED WEEVILS

BELOW | *Cerocranus extremus* From New Caledonia, this beetle has a tower of waxy secretion on its thorax. The purpose of this removable structure is not known.

The Curculioninae are the largest subfamily in the weevil family Curculionidae, and are found throughout the world. They almost all feed on living plants, and the vast majority are host-specific, that is they are associated with only a single plant genus or species. Furthermore, they often feed on only a particular structure of the plant, so a given plant species may be host to several weevil species. This has driven their rapid evolution, which probably happened at the same time as and together with the diversification of flowering plants at the end of the Cretaceous. Curculioninae are called long-nosed weevils because most species have an elongated rostrum with the biting mandibles placed at the end. This is longer in the females, which use it as a drill to make a hole in the plant substrate into which they lay their eggs using a telescopic ovipositor or egg-laying tube,

SUBFAMILY
Curculioninae

KNOWN SPECIES
23,500

DISTRIBUTION
Worldwide except Antarctica

HABITAT
Forests, swamps, plains, agricultural fields—anywhere that plants grow

SIZE
2–50 mm

DIET
Living plants. Adults feed on the leaves and stems, and larvae develop within the plant itself

NOTES
Some Curculioninae are major pests of agriculture. An example is the Cotton Boll Weevil *Anthonomus grandis*, which crippled the cotton industry in the American South. There are also a very few exceptions that don't feed on living plants at all. Members of the European genus *Archarius* develop in

generally the same length as the rostrum. The antennae are angled (geniculate) and the first segment is withdrawn into a groove during drilling, to protect the antennae from harm. The female lays a single egg in each hole, and the larvae that hatch from the egg are white legless grubs, which develop surrounded by food and protected from desiccation, predators, and parasitoids. Numerous species, such

ABOVE | *Curculio nucum* The Hazelnut Weevil drills into hazelnuts and lays one egg in each nut. It can be a minor pest of nut agriculture.

as those in the genus *Curculio*, lay their eggs in nuts or seeds, and the larvae grow within the protection of the seed's shell.

RIGHT | *Erodiscus* This Pinocchio Weevil from South America has one of the longest rostra of any weevil relative to its body length.

galls made by wasps on the leaves of the host plants, and the South American *Ludovix fasciatus* oviposits (lays its eggs) in the egg cases of the grasshopper *Cornops aquaticum* (Orthoptera), one of the few cases of carnivory in the weevils

BROAD-NOSED WEEVILS

The large subfamily Entiminae are usually called broad-nosed weevils because, unlike most other weevils, the rostrum (informally called a "nose," though it is really an extension of the head with the jaws at the end) is short and stout, housing the powerful muscles for the large biting mandibles.

Long-lived adult entimines are able to consume a wide variety of plant matter, including very tough leaves, twigs, and the needles of conifers. Unlike most weevils, they are usually polyphagous, meaning they are not restricted to any particular plant family. The larvae, white legless grubs with a brown head, live in the soil, eating roots, and also feed on a range of different plant families. Some species are pests of horticulture, forestry, or agriculture, chewing notches out of the leaves of crops, trees, or garden plants, while their larvae eat the roots underground. The adults are quite robust and long-lived, and many

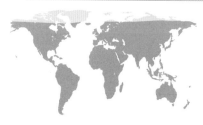

SUBORDER
Entiminae

KNOWN SPECIES
12,000

DISTRIBUTION
Worldwide except Antarctica, although several species are found in the sub-Antarctic islands

HABITAT
Forests, plains, gardens, almost anywhere where plants grow

SIZE
2–25 mm

DIET
Larvae feed in the soil on a variety of roots, while adults feed on the green parts of the plant, especially the leaves

NOTES
The metallic scales of entimine weevils have a structural color, which means it is caused not by pigments but by refraction and reflection of light, working in the same way as some

ABOVE | *Pachyrhynchus congestus pavonius* Beetles in this hard-bodied, flightless genus occur in the Philippine region, and may disperse between islands, carried on seaborne debris.

LEFT | *Entimus granulatus* From South America, these magnificent colorful beetles are sometimes used to make jewelry.

OPPOSITE | *Rhinoscapha*, from New Guinea. The lower leaf in the picture shows the characteristic "notching" caused by adult entimine feeding.

precious stones, such as opals. This evolved to make the beetles difficult to see in the dappled shade of a tropical forest, but because it is structural it doesn't fade, even after the insect's death. Hence, specimens in collections are as bright as the day they were collected. For this reason, parts of the exoskeleton of some entimines are used in jewelry and traditional costume

genera have lost the ability to fly. In some instances, the elytra are fused together, a strategy to conserve water by preventing evaporation, which has allowed entimines to colonize very dry environments. Another strategy that has aided the success of some entimine genera is the evolution of parthenogenesis, where the females can reproduce asexually, effectively cloning themselves. This allows populations of some pest species such as the Black Vine Weevil *Otiorhynchus sulcatus* to grow very quickly.

Many tropical entimines, like those illustrated here, are extremely colorful, the color being formed by hundreds of tiny, metallic reflective scales.

BARK BEETLES

Scolytinae, bark beetles, is a weevil subfamily that lost the characteristic weevil rostrum and became wood borers. Bark beetles live between the wood and the bark of coniferous or deciduous trees. Most are host-specific, attacking only one genus or species of tree. A few are polyphagous, which means they feed on a broad range of hosts, and even fewer feed on herbaceous plants, especially legumes (Fabaceae).

Many species excavate intricate tunnels under the bark, with tunnel patterns identifiable to genus or species. These often consist of a central tunnel, excavated by the female, and smaller tunnels radiating away from it, where her eggs have hatched and the larvae dig their own tunnels, growing in the process. In many species, the female has pocketlike organs called mycangia from which she introduces spores of fungi, which then grow and provide food for the larvae.

Some of these fungi are pathogenic to the trees, so some bark beetles are vectors of tree diseases, such as the genus *Scolytus*, which spreads Dutch elm disease *Ophiostoma novo-ulmi*, a fungus that killed millions of mature elms across Europe, Asia, and North America in the late twentieth century. Some elms remain, but when they reach a certain size, they are infested by the beetles and killed by the fungus, so the gigantic elms that form the background to many eighteenth-century paintings have disappeared from the landscape. Other bark beetles, of the genus *Dendroctonus*, attack climate-stressed conifers in North America, and the results of outbreaks are discussed in the section on forestry (see page 58).

SUBFAMILY
Scolytinae

KNOWN SPECIES
6,000

DISTRIBUTION
Worldwide except Antarctica

HABITAT
Forests, anywhere with trees. Only a few species feed on herbaceous plants

SIZE
1–9 mm

DIET
Wood and wood-feeding fungi

NOTES
Scolytinae are now recognized as true weevils, but for many years they were classified as a separate family of their own because they lack the beaklike rostrum, the projection of the front of the head with the mandibles at its apex, which was considered a defining weevil feature. The rostrum is used by weevils as a drill to prepare a hole for egg laying, but in

ABOVE | *Ips typographus* The Engraver Beetle is a major pest of spruce, including Christmas trees, in much of Europe and northern Asia.

OPPOSITE | *Cnestus mutilatus* The strange Camphor or Sweetgum Shot Borer, an invasive species from Asia, has established in the USA.

RIGHT | *Hylesinus crenatus* The Large Ash Bark Beetle from Europe. This particular specimen was photographed in Switzerland.

Scolytinae there is no longer any need for a rostrum because the whole body has become a drill, with teeth on the thorax in many species being used to rasp away at the wood in order to bore tunnels. A Scolytinae infestation that is recent can be recognized by the fresh sawdust thrown out of their excavations

GLOSSARY

Anthropocentric: worldview centered around people and their needs.

Batesian mimicry: form of imitation where a harmless animal resembles one that is dangerous or toxic, thereby gaining protection from the resemblance. Named after the Victorian naturalist Henry Walter Bates (1825–92), who discovered this process while studying butterflies in the Amazonian rainforests.

Cerrado: tropical dry savannah of South America, covering around a fifth of Brazil and areas of Bolivia and Paraguay.

Chaparral: scrub habitat with hot, dry summers and stunted, drought-tolerant vegetation, characteristic of southern USA and northern Mexico.

Conglobation: rolling up into a ball shape.

Cosmotropical: animal or plant that occurs throughout, or widely in, the tropical regions of the world.

Coxal plates: expanded plate-like structures found on the underside of some adult beetles behind the legs. Useful for taxonomy and identification.

Crypsis: ability to escape detection by blending into the surrounding area, using color, shape, or pattern.

Diurnal: active during daylight.

DNA: deoxyribonucleic acid, the molecule that carries genetic information in both animals and plants. Composed of four chemicals called nucleotides, and their order, or sequence, can be interpreted and of taxonomic importance.

Ectoparasites: organisms that live on the outside of another organism, the host, and take nutrition from it (for example, by sucking blood or sap). Often adapted to live among fur, feathers, for instance, and may be specific as to the species of host that they use.

Elytra: wing cases of beetles, developed from the first pair of wings. At rest, they protect the delicate flight wings.

Flabellate (of antennae): forming plates, or flabellae, which seek to maximize surface area, usually for scent detecting.

Fossorial (of legs): adapted for digging.

Garrigue: scrub habitat of the Mediterranean, consisting of stunted trees and drought-tolerant, aromatic vegetation.

Geniculate (of antennae): knee-shaped (that is, with one part at a marked angle to the other).

Holarctic: occurring throughout the north of the northern hemisphere (that is, naturally occurring in northern Asia, North America, and northern Europe).

Hygropetric: occurring on rock surfaces that are constantly wet.

Inquiline: occurring within the nests of social insects such as ants or termites, using the nest for food or shelter, and having a varying degree of protection from the host insects.

Interstitial fauna: found in caves, but also in the fissures and cracks in the ground that connects caves together.

Lamellate (of antennae): resembling the pages of a book, developed into plate-like lamellae to increase surface area, usually for scent detecting, but differing from *flabellate* antennae by being mobile and able to be opened or closed.

Larviform: resembling a larva (that is lacking wings), even when adult, such as glowworms (Lampyridae).

Maquis: dense Mediterranean coastal scrub habitat, resembling *garrigue*.

Mesonotum: upper part of the mesothorax in adult beetles, where the elytra are attached. Usually covered by the pronotum and elytra when at rest.

Mesoventrite: lower part of the *mesothorax*, where the middle legs are attached.

Mesothorax: middle part of the thorax, including the *mesonotum* and *mesoventrite*.

Metanotum: upper part of the *metathorax* in adult beetles, where the flight wings are attached. Usually covered by the *elytra* when at rest.

Metathorax: third and last part of the thorax, including the *metanotum* and metaventrite.

Mutualism: relationship between two organisms of different species, where both benefit.

Mycangia: cavities in the exoskeleton of some adult beetles (such as Curculionidae: Scolytinae) for carrying fungal spores, with which they inoculate plants to provide food for themselves or their larvae. Some such beetles are vectors of fungal diseases of trees.

Oviposition: laying eggs. Female beetles have an internal ovipositor (egg-laying tube).

Palearctic region: biogeographic region that covers Europe; western, central, and northern Asia; and northern Africa.

Parasitoid: organism that develops on or in another organism, but unlike a true parasite, it kills the host during development.

Parthenogenesis: nonsexual reproduction. Parthenogenetic insects such as vine weevils (genus *Otiorhynchus*) lay eggs without needing a male. These eggs hatch into genetic clones of their mother.

Pectinate (of antennae): developed into a series of long structures like the teeth of a comb.

Pedogenetic: insect able to reproduce while still in the larval stage, such as *Micromalthus* (Archostemata).

Phoretic: obtaining transport (but not necessarily nutrition) by attaching to another organism (such as phoretic mites on beetles).

Physogastric (of some female beetles): with the abdomen able to distend greatly, especially when carrying a large number of eggs.

Phytophagous: eating leaves or living plant tissue, herbivorous.

Plastron: series of tiny setae that maintain a permanent air bubble, allowing an insect to breathe under water without having to renew its air supply at the surface.

Polyphagous: feeding on many different foods (usually in beetles on many different genera or species of plants).

Pronotum: upper part of the prothorax; in an adult beetle the only part of the thorax visible from above between the head and elytra. Often simply called "the thorax," but this is inaccurate.

Prosternum: lower part of the prothorax, to which the front legs are attached.

Quinones: complex organic compounds, used by some beetles for defense.

Relict/Relictual: left behind, or appearing as if left behind, from a previous time. Seemingly primitive in form or appearance.

Rostrum: extension of an insect's head in front of the eyes and with the mouthparts at the end. An example is the "beak" or "snout" of weevils (Curculionoidea).

Saprophagous: feeding on dead or decaying matter.

Scutellum: triangular or shield-shaped piece of the thorax visible from above in some adult beetles, behind the pronotum and between the bases of the elytra.

Setae (sing. **seta**): bristles, or hairs, on an insect's exoskeleton.

Sister Group: in phylogenetics, the closest recognized equivalent-rank relative to a given organism (for example, Sister Species, Sister Genus).

Spiracles: the breathing holes in the exoskeleton of an insect.

Structural color: color that is formed not by pigment but by scattering of light by thousands or millions of microscopic prisms. Found in some beetles, butterfly wings, for example.

Subcortical: living underneath the bark of trees.

Synanthropic: living with, or associated with, humans or human-modified areas.

Tarsi: an insect's "foot," the last part of the leg, usually divided into three to five segments called "tarsomeres." The claw-bearing segment is referred to as the "pretarsus."

Taxonomy: the science of classification of living organisms.

Temperate zones: regions of the world between the poles and the tropics, usually marked by four distinct and clearly recognizable seasons.

Tergites: segments of the abdomen.

Troglophile: organism associated with and able to live its entire life in a cave.

Xerophilous: inhabiting and able to tolerate dry or very dry environments.

Xerophytic (of plants): growing in very dry habitats, usually with adaptations to conserve water, such as small leaves and aromatic chemical oils.

RESOURCES

SELECTED BOOKS AND FIELD GUIDES

Albouy, V. and D. Richard. 2017.
Coléoptères d'Europe.
Delachaux et Niestlé.

Arnett, R. H. Jr. and M. C. Thomas (Eds). 2001.
*American Beetles. Volume 1. Archostemata, Myxophaga,
Adephaga, Polyphaga: Staphyliniformia.*
CRC Press.

Arnett, R. H. Jr., M. C. Thomas,
P. E. Skelley and J. H. Frank (Eds). 2002.
*American Beetles. Volume 2. Polyphaga: Scarabaeoidea
through Curculionoidea.* CRC Press.

Bouchard, P. (Ed.). 2014.
*The Book of Beetles: A Life-size Guide to Six Hundred
of Nature's Gems.* The University of Chicago Press.

Cooter, J. and M. V. L. Barclay (Eds). 2006.
A Coleopterist's Handbook (4th edition).
Amateur Entomologists' Society.

Dodd, A. 2016.
Beetle.
Reaktion Books Ltd.

Durrell, G. and L. Durrell. 1982.
*The Amateur Naturalist: A Practical Guide to the Natural
World.* Hamish Hamilton Ltd., UK.

Evans, A. V. 2014.
Beetles of Eastern North America.
Princeton University Press.

Evans, A. V. and C. L. Bellamy. 1996.
An Inordinate Fondness for Beetles.
Henry Holt & Co. Inc.

Evans, A. V. 2021.
Beetles of Western North America.
Princeton University Press.

Hangay, G. and P. Zborowski. 2010.
A Guide to the Beetles of Australia.
CSIRO.

Harde, K. W. and F. Severa. 1984.
A Field Guide in Colour to Beetles.
Octopus Books, London.

Jones, R. 2018.
Beetles: New Naturalist Series.
Harper Collins.

Lawrence, J. F. and A. Ślipiński. 2013.
*Australian Beetles. Volume 1. Morphology,
Classification and Keys.*
CSIRO.

Leonard, M. G. 2018. *The Beetle Collector's Handbook.*
Scholastic, UK.

Marshall, S. A. 2018.
Beetles: The Natural History and Diversity of Coleoptera.
Firefly Books Ltd.

Ślipiński, A. and J. F. Lawrence. 2019.
*Australian Beetles. Volume 2. Archostemata,
Myxophaga, Adephaga, Polyphaga (part).*
CSIRO.

ONLINE RESOURCES

Asociación Europea de Coleopterologia [Spain]
uia.org/s/or/en/1100046855

BugGuide
bugguide.net/node/view/60

Coleopterological Society of Japan
www.kochugakkai.sakura.ne.jp/English/index2.html

iNaturalist
www.inaturalist.org/taxa/47208-Coleoptera

The Coleopterist Journal
www.coleoptera.org.uk/coleopterist/home

The Coleopterists Society
www.coleopsoc.org

Wiener Coleopterologen Verein [Austria]
www.coleoptera.at

ABOUT THE AUTHORS

Maxwell V. L. Barclay is President of the Coleopterists Society and Senior Curator of Coleoptera (Beetles) at the Natural History Museum, in London, UK. He manages a team of expert staff and one of the largest, oldest, and most comprehensive beetle collections in the world, including 8 million beetle specimens of around 250,000 species. A passionate entomologist, Max strives to increase awareness and knowledge of entomology, insect collections, and the natural habitats that sustain us all.

Patrice Bouchard is a scientist at the Canadian National Collection of Insects, Arachnids, and Nematodes. His research focuses on the systematics of weevils and darkling beetles, with a focus on groups of importance to agro-ecosystems. He is active in scientific societies such as the Entomological Society of Canada and the International Commission on Zoological Nomenclature. Other co-authored books include *Tenebrionid Beetles of Australia* and *The Book of Beetles*.

INDEX

A

Abax 98
abdomen 8, 10, 16–17
Adephaga 90–107
Aderidae 43
Agrilinae 146–7
Agrilus 59, 146–7
Agriotes 157
Agrypninae 158–9
Agyrtidae 117
Alaobia 125
Aleochara 125
Aleocharinae 41, 124–5
Amarygmus 193
amber 22–3, 27
Ampedus 156–7
Amphizoa 48
Amphizoidae 48, 93
Anning, Mary 74
Anobiinae 168
Anobium 61
Anomala 140
Anoplognathus 141
Anoplophora 59, 204–5
Anorus 145
ant nests 40–1, 114, 127
ant-like flower beetles 198–9
ant-like litter beetles 126–7
antennae 13, 14–15, 19
Anthicidae 198–9
Anthonomus 55
Anthrenus 60–1
Anthribidae 216–17
Anthribus 217
antimicrobials 77
aphids 25, 185
Aphodiinae 38
Apioninae 220–1
aquatic beetles 12, 48–51,
  87–9, 92–3, 102–7, 112–13,
  152–3

Araecerus 216
Archostemata 82–5
art 68–9
Asian Longhorn 59, 204–5
Aspidytidae 92–3
Astylus 176
Atlas Beetles 18, 129
Attelabidae 218–19
Attenborough, Sir David 75
augur beetles 168–9

B

Banks, Joseph 53
Baranowskiella 116
Baridinae 14
bark beetles
  30, 59, 172, 174, 179, 226–7
bark-gnawing beetles 172
Bates, Henry Walter 74, 202
beaches 46–7
Beaver Beetles 118–19
"Beetle Wing Dress" 69
Belohina 128
Belohinidae 128
biological control
  35, 64–5, 166, 207, 221
bioluminescence
  15, 16–17, 43, 154, 159, 162–3
biomimetics 76–7
biospeleology 44–5
Biphyllidae 173
Blaps 61, 79, 192
Bledius 101
blister beetles 78, 196–7
Bolboceratinae 131
bombardier beetles 17, 91
Bostrichidae 168–9
Bostrichoidea 168–9
Brachininae 91
Brachinus 17
Brentidae 220–1

Brentinae 220
broad-nosed weevils 224–5
Bruchinae 19, 54
bubble bugs 78
Buprestidae 12, 59, 69, 146–51
Buprestinae 148–9
Buprestis 29, 52, 148
burrowing ground beetles
  100–1
burying beetles 33, 39, 77, 120
Byrrhoidea 152–3

C

Callirhipidae 153
Callirhipis 153
Cantharidae 164–5
cantharidin 78, 197
capricorn beetles 206–7
Carabidae 13, 17, 24–5, 49, 50,
  91, 92, 96–101
carpet beetles 60
Carpophilus 183
carrion beetles
  32–3, 118, 120–1
Cassidinae 208–9
caves 44–5
cedar beetles 153
cellar beetles 61, 192
Cerambycidae 58–9, 202–7
Cerambycinae 206–7
Ceratocanthinae 134–5
Cetoniinae 129
Chalcosoma 18–19, 129
checkered beetles 172–5
Cholevinae 118
Christmas beetles 141
Chrysina 141
Chrysochroinae 150–1
Chrysomelidae 16, 19, 26,
  54–5, 64, 69, 208–13
Chrysomeloidea 200–1

cicada parasite beetles 144–5
Cicindela 52, 96
Cicindelinae 12, 25, 49, 96–7
Ciidae 186
Clambidae 142–3
classification 80–1
Cleridae 174–5
Cleroidea 172–3
click beetles 15, 43, 154–9
clown beetles 110, 114–15
Coccinella 15
Coccinellidae
    15, 25, 79, 178–9, 184–5
Colophon 133
Colorado Potato Beetles 54–5
complete metamorphosis
    9, 20–1
conglobation 135
Coprophanaeus 33
Cotton Boll Weevil 55
crawling water beetles 102–3
Crowsoniellidae 83
Cryptocephalinae 210–11
Cucujoidea 178–9
Cupedidae 83, 84–5
Curculionidae
    26, 55, 58–9, 214–15, 222–7
Curculioninae 222–3
Curculionoidea
    13, 64–5, 66–7, 214–27
Cylas 221
Cylindrical Bark Beetles
    188–9

D

darkling beetles
    186, 190, 192–5
Darwin, Charles 7, 18, 53, 74
Dascillidae 144–5
Dascilloidea 144–5
Dascillus 145

Death Watch Beetles 61, 168
decaying animals 32–3
decaying plants 28–9
Decliniidae 142, 143
defensive chemicals
    17, 25, 26–7, 73, 77, 78, 91,
    107, 123, 160, 180, 193, 197
Dejean, Pierre 53
Dendroctonus 59, 179, 226
Dendroxena 120
Dermestes 168
Dermestidae 12, 33, 60, 168
Dermolepida 65
Derodontidae 166–7
Derodontoidea 166–7
desert beetles
    46–7, 73, 77, 190–1
Devil's Coach Horse
    17, 25, 123
Dinapate 168
diseased-tree beetles 166, 167
Disteniidae 200, 201
diving beetles
    25, 92–3, 104–7
Dor Beetles 130–1
dragon bugs 78
driftwood 46
drought tolerance
    46–7, 190, 225
Dryopoidea 152
dung beetles
    29, 33–5, 37, 38, 68, 72, 130–1,
    136–7
Dürer, Albrecht 68
Durrell, Gerald 76
Dutch elm disease 30, 59, 226
Dynastes 14, 29, 129, 139
Dynastinae 129, 138–9
Dyschirius 101
Dytiscidae 25, 104–7
Dytiscinae 104–5

E

Earth-boring Dung Beetles
    130–1
Elaeidobius 63
Elateridae 15, 43, 156–9
Elaterinae 156–7
Elateroidea 19, 154–9, 165
Eleodes 73
Elephant Beetles 129
elm bark beetles 30, 59
Elmidae 48, 152
elytra 8, 10, 14, 16, 22, 68–9
Emerald Ash Borer 59, 147
Emmita 145
Endecatomidae 168
Entiminae 224–5
entomophagy 66–7
Epilachna 179
Epilachninae 185
Erotylidae 180–1
Euchroma 151
Eucinetidae 142, 143
Eulichadidae 153
Eurynebria 50–1
Exapion 64–5
exoskeleton 14, 21, 22
eyes 12, 13, 44, 94–5

F

Fabre, Jan 69
featherwing beetles 117
fireflies 14, 78, 79, 154, 162–3
flat-faced longhorn beetles
    204–5
flat-headed jewel beetles
    146–7
flea beetles 212–13
flightlessness
    102, 144, 145, 191
flower longhorns 202–3
flower-feeding beetles 12

fog basking 47, 77
food for humans 66–7, 158
food webs 36–7
forest stream beetles 153
fossils 22–3, 41, 84, 87, 91, 136
freshwater species 48–9
fungi 30–1, 180–1, 193
fungus weevils 216–17

G

Galerucinae 212–13
Gastrophysa 16
genitalia 16
Geotrupes 71
Geotrupidae 130–1
Giant Acacia Click Beetles 158
Giant Ceiba Borer 151
Giant Longhorn 29
Giant Palm Borer 168
glowworms 154, 162–3
Goethe, 80
Goliath Beetles 129
Goliathus 129
Gorse Weevil 64–5
Gray, Thomas 70
great diving beetles 104–5
ground beetles
   24–5, 92–3, 96–101
Gyrinidae 12, 79, 91, 94–5

H

Haldane, J. B. S. 6
Haliplidae 102–3
Haliplus 103
ham beetles 174
Harmonia 79
harmonine 77, 79
Harpalinae 98–9
Harpalus 98
head 8, 10, 12–13
Heliocopris 136

Helophorus 113
Hemimetabola 9, 21
Hercules Beetles
   14, 129, 139
hide beetles 12, 168
hister beetles 110, 114–15
Histeridae 110, 114–15
Holometabola 9, 20–1
horns 13, 14, 18
humans
   association 52–79
   food 66–7, 158
Hybosoridae 134–5
Hybosorinae 134–5
Hydraenidae 117
Hydrophilidae 110, 112–13
Hydrophiloidea 110–11
Hydroporinae 106–7
Hydroporus 107
Hydroscaphidae 87, 88
Hygrobiidae 93
Hypocephalus 200

I

ironclad beetles 188–9
Ischalia 199
Ischaliidae 199
Ithycerinae 221

J

Jacobsoniidae 166, 167
Japanese beetles 140
jewel beetles
   12, 29, 69, 141, 146–51
Julodimorpha 148
Jurodidae 83

K

Kafka, Franz 71
Keats, John 70
Khepri 72

L

ladybugs
   15, 25, 31, 77, 79, 178, 184–5
Lamiinae 204–5
Lampyridae 14, 16–17, 154,
   162–3
Lampyris 162–3
Languriinae 181
larder beetles 60
large jewel beetles 150–1
Laricobius 166
leaf beetles
   16, 26, 64, 200–1, 208–13
leaf-cutter ants 41
leaf-miners 146
leaf-rolling weevils 218–19
Leiodidae 118–19
Leiodinae 118
Leonard, M. G. 71
Lepiceridae 87, 89
Lepiceroidea 87
Lepicerus 89
Leptinotarsa 54–5
Lepturinae 202–3
Lethrinae 131
lightning bugs 162–3
Linnaeus, Carolus 52–3, 80
literature 70–1
Lizard Beetles 181
"lock and key" hypothesis 16
long-nosed weeviles 222–3
longhorn beetles
   29, 59, 200–5
Lucanidae
   13, 19, 81, 132–3
luciferase 78, 162
Luciola 79
Lycidae 160–1
Lymexylidae 170–1
Lymexyloidea 170–1
Lytta 78, 197

M

Malachiinae 176

Malacoderms 176

Mamboicus 101

mandibles 12–13

Mantell, Gideon and Mary 74

Manticora 97

marsh beetles 142–3

mate attraction 18, 162–3

mating 16, 18–19

mealworms 67, 78–9, 192

medicine 78–9

Megacephala 97

Megadytes 104

Megalopodidae 200, 201

Megaloxantha 151

Megasoma 29, 129, 138

Megaxenus 43

Melanophila 148–9

Meligethes 182–3

Meloe 196–7

Meloidae 78, 196–7

Melyridae 176–7

Meruidae 92

metallic darkling beetles 194–5

Mexican Bean Beetles 179

Micromalthidae 83, 84

Mimela 140

Molliberus 165

Monotomidae 179

Mountain Pine Beetles 59

museum beetles 60–1, 168

museums 74

use of beetles 33

mycangia 30, 170, 226

Myrmeconycha 212

myrmecophiles 41

mythology 72–3

Myxophaga 86–9

N

Nairobi Eye Fly 78

Necrobia 174

Necrodes 120

nests 38–43

ants 40–1, 114, 127

termites 42–3, 114, 158

vertebrate 38–9

net-winged beetles 160–1

Nicrophorinae 120

Nicrophorus 33, 39, 77, 120

Nitidulidae 178, 182–3

Nosodendridae 166, 167

Notoxinae 199

O

ocelli 12

Ochyropus 101

Ocypus 17, 25, 123

Oedemeridae 46, 78

oil beetles 196–7

Ommatidae 83, 84–5

Omorgus 39

Ontholestes 12

Onymacris 47

Orchid Beetles 145

Oryctes 139

Oxypeltidae 200, 201

P

Paederus 78

Palaeosastra 212

Palm Rhinoceros 139

palm weevils 67

parasites

of beetles 31

beetles as 39, 174

of humans 79

pederin 78

pedogenetic larvae 84

pesticides 37

pests

crops 54–5, 98–9, 129, 176–7, 179, 182–3, 219, 221, 224

domestic 60–1, 174

gardens 60, 140, 224

stored produce 56–7, 216

timber 58–9, 170, 205, 207, 224

truffle farms 131

Phaeochroops 134–5

Phaeochrous 134–5

Phengodidae 154

pheromones 13

Phoracantha 206–7

phoretic mites 32–3

Phosphuga 120

Photinus 163

Photuris 163

physogastric species 16

Pimeliinae 190–1

plant feeders 26–9, 54–5

plastron 48

Platerodrilus 161

Platypsyllinae 118–19

Platypsyllus 119

pleasing fungus beetles 180–1

Pokémon 74

pollen beetles 182–3

pollination 36, 62–3, 176, 182, 202, 211

Polychalca 69

Polyphaga 108–227

Popillia 140

popular culture 74–5

pot beetles 210–11

predation 24–5

by beetles 46, 50

of beetles 36–7

Priacma 85

primitive weevils 220–1

pronotum 10, 14, 15

Prosopocoilus 133
Protocoleoptera 22
Psammodes 191
Pselaphinae 41, 126–7
Psephenidae 48
Pseudovadonia 203
Pterostichus 98
Ptiliidae 116–17
Ptinidae 61, 168
Ptininae 168
pygidium 16
Pyrearinus 43, 158–9
Pyrophorus 159

R

Reesa 60–1
reflex bleeding 180
respiration 16
Reticulated Beetles 85
Rhagophthalmidae 154
rhinoceros beetles
  18, 29, 138–9
Rhinocyllus 64
Rhipiceridae 144–5
Rhizophagus 179
Rhynchites 219
Rhynchitinae 218, 219
Rhynchophorus 67
Rhysodidae 93
Riffle Beetles 48, 152
Rosalia 207
rostrum 13, 15
round fungus beetles 118
Rousseau, Jean-Jacques 80
rove beetles 16, 116–17, 122–5
Rutelinae 140–1

S

Sade, Marquis de 78, 197
Sagrinae 19
saltwater species 50–1

sand habitat 46–7
sap beetles 182–3
scarab beetles 34, 38, 66–7, 68,
  72, 128–9, 134–5
Scarabaeidae
  13, 29, 34, 128–9, 136–41
Scarabaeinae 136–7
Scarabaeoidea
  19, 66–7, 81, 128–9, 134–41
Scaritinae 100–1
scavenger scarab beetles
  134–5
Scirtidae 142
Scirtoidea 142–3
Scolytinae 226–7
Scolytus 30, 59
screech beetles 93
Scydosella 116
Seven Spot Ladybird 15
sexual selection 18–19
Shakespeare, William 70–1
shining leaf chafers 140–1
ship timber beetles 170–1
Silphidae
  32–3, 39, 77, 120–1
Silphinae 120
Sinopyrophorus 154
skiff beetles 88
small carrion beetles 118
small diving beetles 106–7
soft-bodied plant beetles
  144–5
soft-winged flower beetles
  176–7
soldier beetles 164–5
sooty bark disease 30
Spanish Fly 78, 197
Sphaeridiinae 113
Sphaeritidae 110–11
Sphaerius 88
Sphaeriusidae 87, 88

Sphaeriusoidea 87
spider beetles 168
spiracles 16
Spotted Maize Beetles 176
stag beetles
  13, 19, 68, 81, 132–3
Staphylinidae 12, 16, 17, 25,
  33, 77, 78, 116, 120, 122–7
Staphylininae 122–3
Staphylinoidea 116–27
Stelidota 183
Stenocara 47
Stenochiinae 194–5
Sternocera 69
Strawberry Beetles 98
Strepsiptera 9
Strongylium 194
sugarcane beetles 65
Sweet Potato Weevil 221
symbiosis 31, 170
synanthropic species 60–1
Synteliidae 110–11

T

Tajiri, Satoshi 74
Tenebrio 67, 78–9, 192
Tenebrionidae
  47, 67, 73, 79, 186, 190–5
Tenebrioninae 192–3
Tenebrionoidea
  9, 186–7, 190–7, 198–9
Tenebroides 172–3
tergites 16
termite nests 42–3, 114, 158
Terry, Ellen 69
Tetralobus 158
Thistle Weevil 64
thorax 8, 10, 14–15
tiger beetles 12, 25, 96–7
Titanus 29
toktokkies 191

tooth-necked fungus beetles 166–7
torrent beetles 88–9
Torridincolidae 87, 88–9
tortoise beetles 69, 208–9
Trachypachidae 93
Trichodes 174
trichomes 41
Trictenotomidae 186
Trogidae 39
troglophiles 44–5
Trogossitidae 172
trout stream beetles 48, 93
twisted-winged parasites 9

U

Ulomoides 79

V

vertebrate nests 38–9
Vesperidae 200

W

Wallace, Alfred Russel 7, 8, 53, 74
Water Penny Beetles 48
water scavenger beetles 110, 112–13
weevils 13, 14, 15, 26, 63, 64–5, 66–7, 214–27
whirligig beetles 12, 79, 94–5
Wilson, E. O. 37
wings 14, 16
wireworms 157
wood-boring beetles 46, 114, 170–1, 172, 174, 226–7
Woodworm 61, 168
woolly bears 168
Wordsworth, William 71

X

Xestobium 61

Y

Yellow Mealworm 78–9

Z

Zabrus 99
Zopheridae 188–9
Zopherus 188–9

PICTURE CREDITS

The authors and publisher would like to thank the following for permission to reproduce copyright material. All reasonable efforts have been made to trace copyright holders and to obtain their permission for the use of copyright material. The publisher apologizes for any errors or omissions and will gratefully incorporate any corrections in future reprints if notified.

l=left; r=right; t=top; b=bottom; m=middle.

2–3, 6–7, 8, 9 Kenji Kohiyama; 5 Dreamstime/Tirrasa; 12, 13t, 15t, 18t, 18b, 101t, 108, 133b, 137t, 137b, 140, 149b, 193b, 209t, 216, 218, 223t Eugenijus Kavaliauskas; 13b, 17t, 97t, 135b, 143b, 147t, 153t, 153b, 159t, 159b, 164, 171b, 172, 175t, 181t, 183t, 185t, 186, 194, 195b, 199, 208, 211r, 214, 217t, 219t, 221t Nicky Bay; 14, 174 Hello MuMu; 15b Christos Zoumides; 16, 227b Pierre Bornand; 17b Tony (tickspics); 19, 96, 180, 197t, 238 Nature Picture Library/Piotr Naskrecki; 20l Maria Justamond (rockwolf); 20r, 123t, 161b, 163b, 163t, 165t, 181b, 185b, 210, 213b, 223b, 225b © Andreas Kay, images courtesy of Heinz Schneider and the family and estate of Andreas Kay; 22 Alekseev, V. I., Mitchell, J., McKellar, R. C., Barbi, M., Larsson, H. C. E., and Bukejs, A., "The first described turtle beetles from Eocene Baltic amber, with notes on fossil Chelonariidae (Coleoptera: Byrrhoidea)," Foss. Rec., 24, 19–32, https://doi.org/10.5194/fr-24-19-2021, 2021; 23l, 110 Rixin Jiang; 23r, 166 Chenyang Cai; 24, 120, 150, 225t Frank Deschandol; 25t Thomas Langhans; 25b Warren photographic; 26 Klaus Bolte; 27t, 90, 98, 122, 132, 138, 139b, 160, 179t, 195t, 196, 201t, 205b, 224 Chien C. Lee; 27b Carina Van Steenwinkel; 28, 130, 148, 157b, 169t, 169b, 203, 207b Tamás Németh; 29t Thomas J. Astle; 29b Tom Patterson; 30 M. W. Baker; 31t, 78, 165b, 187b, 197b, 219b, 220 Ryszard Szczygieł; 31b, 217b Carroll Perkins; 32, 145 Joyce Gross; 33 Achim Kluck; 34 iStockphoto/Henrik_L; 35t, 136, 191t, 212 Bernard Dupont; 35b Job Aben, 36t K.P. McFarland–Vermont Atlas of Life; 36–37 Shutterstock/Stu Porter; 38 Jim McClarin; 39t Marlin E. Rice; 39b Dreamstime/Viniciussouza06; 40 Getty Images/Paul Starosta; 41, 191b Christopher C. Wirth; 42 Maruyama M. "Termitotrox cupido sp. n. (Coleoptera, Scarabaeidae), a new termitophilous scarab species from the Indo-Chinese subregion, associated with Hypotermes termites," 2012. ZooKeys, 254: 89–97. Pensoft Publishers. DOI:10.3897/zookeys.254.4285.; 43t, 114, 125b, 135t Pavel Krásenský; 43b Alamy Stock Photo/Minden Pictures; 44 Alex Hyde; 45, 127t Nikola Rahmé; 46 Shutterstock/Chantelle Bosch; 47t, 49b Simon Grove/TMAG; 47b Alamy Stock Photo/blickwinkel/Hecker; 48, 94, 102, 103t, 104, 106, 112, 113t, 142 Jan Hamrsky; 49t Yejie Lin; 50t Wikimedia Commons/Warren Steiner; 50b Derek Sikes; 51t, 51m David Fenwick; 52t Felipe E. Rabanal Gatica; 52b Jean-Yves Rasplus; 54 Shutterstock/Kuttelvaserova Stuchelova; 55t Alamy Stock Photo/Nigel Cattlin; 55b © His Majesty the King in Right of Canada as represented by the Minister of Agriculture and Agri-Food/Christine Noronha; 56 Shutterstock/Tomasz Klejdysz; 57t Dreamstime/Tomasz Klejdysz; 57b, 65b, 119t, 171t Radim Gabriš; 58 Marco Colombo; 59t Alamy Stock Photo/Clarence Holmes Wildlife; 59b Li, Y., Simmons, D. R., Bateman, C. C., Short, D. P. G., Kasson, M.T., et al., "Correction: New Fungus-Insect Symbiosis: Culturing, Molecular, and Histological Methods Determine Saprophytic Polyporales Mutualists of Ambrosiodmus Ambrosia Beetles," 2016. PLOS ONE 11(1): e0147305. https://doi.org/10.1371/journal.pone.0147305; 60 Tim Worfolk; 61t Shutterstock/Stefan Rotter; 61b iStockphoto/NNehring; 62 Alamy Stock Photo/Deborah Vernon; 63t Bruno de Medeiros 2020; 63b, 161t NL Wild Media/Nick Volpe; 64 Paul D. Pratt, United States Department of Agriculture; 65t Alamy Stock Photo/Nigel Cattlin; 66b iStockphoto/Bigpra; 66–67 iStockphoto/ollo; 67t Dreamstime/Arisa Thepbanchornchai; 68, 69b

Getty Museum Collection; 69t Dreamstime /Alexey Fedorenko; 70 Alamy Stock Photo/dominic dibbs; 71t Rare Books, Archives and Special Collections, The University of Melbourne; 71b *Battle of the Beetles* text © M. G. Leonard, Covers © Chicken House 2016, Illustrated by Julia Sardà, designed by Helen Crawford-White, Reproduced with permission of Chicken House Ltd. All rights reserved; 72 Shutterstock/Danny Ye; 73t Robert Harding/ Jurgen & Christine Sohns; 73b Shutterstock/SagePhotography111; 74 Shutterstock/Purino; 75 Alamy Stock Photo/Album; 76t Alamy Stock Photo/John Cancalosi; 76b Alamy Stock Photo/Ann and Steve Toon; 77 Alamy Stock Photo/Clarence Holmes Wildlife; 79t Petr Mückstein (www.bio-foto.com); 79b Terry Priest; 82, 226 Matt Bertone; 84 Julien Vittier & Vincent Nicolas; 85t, 198 Alex Wild/alexanderwild.com; 85b, 115b, 141b, 183b Dash Huang; 86 Jiří Hájek/Natural History Museum—Czech Republic; 88, 144 © His Majesty the King in Right of Canada as represented by the Minister of Agriculture and Agri-Food/Anthony Davies; 89 David Maddison; 92 Katja Schulz; 93, 95t Alamy Stock Photo/Hakan Soderholm; 95b, 107b, 121b, 128, 155t, 182 John and Kendra Abbott; 97b Wikipedia/Alpsdake; 99t, 129b, 151t, 162, 205b, 206 Mikhail M. Omelko; 99b Dr. Alexey Yakovlev; 100 Eddy Wajon; 101b Nature Picture Library/Claudio Contreras; 103b Nikolai Vladimirov; 105t Jiri Lochman/Lochman LT; 105b Shutterstock/Dirk Ercken; 107t Alamy Stock Photo/blickwinkel/Hartl; 111 Rimvydas Kinduris; 113b Max Harhun; 115t, 143t, 154, 158, 188, 189t, 192 Jeffrey P. Gruber; 116 Alexey Polilov; 117t, 117b, 118 Steve Marshall; 119b, 126, 147b, 149t Alamy Stock Photo/Clarence Holmes Wildlife; 121t Sebastián Jiménez López; 123b Dreamstime/Geza Farkas; 124, 179b Melvyn Yeo; 125t Alamy StockPhoto/Nigel Cattlin; 127b Wikimedia Commons/Zi-Wei Yin & Giulio Cuccodoro; 129t Wikimedia Commons/Siga; 131 Alamy Stock Photo/Bob Gibbons; 133t Shutterstock/Anest; 134 Steve & Alison Pearson Airlie Beach; 139t Heinz Rothacher; 141t Julien Touroult; 144t Jean and Fred Hort; 146 Nick Monaghan—lifeunseen.com; 151b Shutterstock/Petr Mückstein; 152 Benjamin Fabian; 155b Alamy Stock Photo/Oliver Thompson-Holmes; 156 Alamy Stock Photo/Nenad Tasic; 157t Andrew Bradford; 167t Bonnie Ott; 167b Gernot Kunz; 168, 201b Alamy Stock Photo/blickwinkel/H. Bellmann/F. Hecker; 170 Julian Hodgson; 173b Dreamstime/Henrikhl; 173t Shutterstock/Tomasz Klejdysz; 175b Shutterstock/Henrik Larsson; 176 Shutterstock/Celso Margraf; 177t Alamy Stock Photo/Peter Yeeles; 177b Frank Vassen; 178, 193t Mok Youn Fai; 184 Alamy Stock Photo/WILDLIFE GmbH; 187t Stefan Verheyen; 189b Matt Bertone; 190 Shutterstock/Dziajda; 200 João P. Burini; 202 Simon Garneau; 204 Alamy Stock Photo/imageBROKER/André Skonieczny; 207t Agefotostock/Zoonar.com/Jakub Mrocek; 209b Shutterstock/Chui Wui Jing; 211l Shutterstock/macrowildlife; 213t Shutterstock/Tomas Vacek; 215 Kristi Ellingsen; 221b Alamy Stock Photo/fishHook Photography; 222 Damien Brouste; 227t Gilles San Martin. **Cover photos: front cover, clockwise from top left:** Shutterstock/Anton Kozyrev; Shutterstock/alslutsky; Shutterstock/Iulian N; Shutterstock/alslutsky; Shutterstock/Anton Kozyrev; above: Shutterstock/Anton Kozyrev; below: Shutterstock/Henrik Larsson; Shutterstock/Anton Kozyrev; Shutterstock/Nicola Dal Zotto; **spine:** Shutterstock/Anton Kozyrev; **back cover:** Shutterstock/Mark Brandon.

LEFT | *Goliathus regius* (Scarabaeidae) A male Giant Goliath Beetle of Tropical Africa climbs among ancient fig roots in a Guinean forest. One of the heaviest flying insects, it is at risk of disappearing along with its rainforest home.

ACKNOWLEDGMENTS

The authors would like to thank the staff at Bright Press for their guidance and sustained support throughout the project. Editors and anonymous reviewers improved the overall quality of the book. We thank our respective institutions; without their support the production of this book would not have been possible. The following colleagues are thanked for assisting with the identification of beetles in photographs and for helpful comments on the text: A. Ballerio, A. Brunke, H. Douglas, M. Geiser, S. Nikolaeva, and D. Telnov. Digital enhancement of older illustrations was provided by A. Davies. Information regarding beetles in popular culture was provided by J. Lewis. A number of beetle photographs were acquired with the assistance of A. Bennett, J. Girón, V. Grebennikov, J. Hulcr, I. Miko, P. Nagel, and J. Parker. We thank L. Packer and A. Evans for their insight regarding the production of insect-focused books. We express gratitude toward all the photographers who contributed images for this book. Finally, we would like to thank our families for their persistent encouragement.